United States Government Accountability Office

GAO

Report to the Chairman, Committee on Environment and Public Works, U.S. Senate

I0470834

May 2012

NANOTECHNOLOGY

Improved Performance Information Needed for Environmental, Health, and Safety Research

GAO
Accountability ★ Integrity ★ Reliability

GAO-12-427

NANOTECHNOLOGY

Improved Performance Information Needed for Environmental, Health, and Safety Research

G A O
Accountability * Integrity * Reliability

Highlights

Highlights of GAO-12-427, a report to the Chairman, Committee on Environment and Public Works, U.S. Senate

Why GAO Did This Study

Nanotechnology involves the ability to control matter at approximately 1 to 100 nanometers. Worldwide trends suggest that products that rely on nanotechnology will be a $3 trillion market by 2020. However, some of the EHS impacts of nanotechnology are unknown. The NSTC coordinates and oversees the NNI, an interagency program that, among other things, develops national strategy documents for federal efforts in nanotechnology.

In this context, GAO examined: (1) changes in federal funding for nanotechnology EHS research from fiscal years 2006 to 2010; (2) the nanomaterials that NNI member agencies' EHS research focused on in fiscal year 2010; (3) the extent to which NNI member agencies collaborate with stakeholders on this research and related strategies; and (4) the extent to which NNI strategy documents address desirable characteristics of national strategies. GAO's review included seven NNI agencies that funded 93 percent of the EHS research dollars in fiscal year 2010. This report is based on analysis of NNI and agency documents and responses to a questionnaire of nonfederal stakeholders.

What GAO Recommends

GAO recommends that the Director of the Office of Science and Technology Policy (OSTP), which administers the NSTC, (1) coordinate development of performance information for NNI EHS research needs and publicly report this information; and (2) estimate the costs and resources necessary to meet the research needs. OSTP and the seven included agencies neither agreed nor disagreed with the recommendations.

View GAO-12-427. For more information, contact Frank Rusco at (202) 512-3841 or ruscof@gao.gov.

What GAO Found

From fiscal years 2006 to 2010, the National Science and Technology Council (NSTC) reported more than a doubling of National Nanotechnology Initiative (NNI) member agencies' funding for nanotechnology environmental, health, and safety (EHS) research—from approximately $38 million to $90 million. Reported EHS research funding also rose as a percentage of total nanotechnology funding over the same period, ending at about 5 percent in 2010. However, GAO identified several reporting problems that raise concerns about the quality of EHS funding data reported. For example, for 18 percent of the 2010 projects GAO reviewed that were reported as EHS research, it was not clear that the projects were primarily directed at EHS risks. In addition, NNI member agencies did not always report funding using comparable data. The absence of detailed guidance on how agencies should report funding for their nanotechnology research has contributed to these problems, as GAO also reported in 2008 and made a related recommendation.

In 2010, EHS research at the NNI member agencies GAO reviewed most frequently focused on carbon nanotubes, nanosilver, and nanoscale titanium dioxide. NNI has not prioritized nanomaterials for EHS research, but NNI's 2011 EHS research strategy outlines criteria for NNI member agencies to use in doing so. It is too soon to tell how these criteria will influence NNI member agencies' decisions about which nanomaterials to prioritize, and it is unclear if information needed to use the NNI criteria is available.

The NNI member agencies have collaborated extensively on EHS research and strategies. They have collaborated through the NSTC to develop joint EHS research strategies and have initiated numerous formal collaborative EHS research projects. Nonfederal stakeholders who responded to GAO's web-based questionnaire on nanotechnology EHS research told GAO that they benefited from collaboration with the NNI member agencies but identified some challenges, including a lack of funding and limited awareness of collaboration opportunities, among others. Most respondents rated the 2011 NNI EHS research strategy as somewhat or very effective at addressing nanotechnology EHS research needs.

NNI strategy documents for EHS research issued by the NSTC address two and partially address the other four of the six desirable characteristics of national strategies identified by GAO that offer a management tool to help ensure accountability and more effective results. For example, the NNI strategy documents provide a clear statement of purpose, define key terms, and discuss the quality of currently available data, among others. However, they do not include performance information—such as performance measures, targets, and time frames for meeting those measures—that would allow stakeholders to evaluate progress towards the goals and research needs of the NNI. In addition, the documents do not include, or sufficiently describe, estimates of the costs and resources needed for the strategy. Without this information, it may be difficult for agencies and stakeholders to implement the strategy and report on progress toward achieving the research needs and assess if investments are commensurate with costs of the identified needs.

_____ United States Government Accountability Office

Contents

Letter		1
	Background	7
	Funding of EHS Research by NNI Member Agencies Has Increased, but Quality of Funding Data Is Uncertain	15
	EHS Research Funded in 2010 Focused On Metal- and Carbon-Based Nanomaterials	25
	NNI Member Agencies Have Collaborated Extensively on Nanotechnology EHS Research	31
	NNI Strategy Documents Address or Partially Address Desirable Characteristics of National Strategies	43
	Conclusion	50
	Recommendations for Executive Action	51
	Agency Comments and Our Evaluation	52
Appendix I	Objectives, Scope, and Methodology	53
Appendix II	Collaborative Nanotechnology Environmental, Health, and Safety Research Agreements	65
Appendix III	GAO Contact and Staff Acknowledgments	78

Tables

Table 1: Program Component Areas	13
Table 2: GAO Analysis of Nanotechnology Projects Reported by Selected Agencies as EHS Research, Fiscal Year 2010	19
Table 3: Summary of Desirable Characteristics for a National Strategy	43
Table 4: Extent NNI Strategy Documents Address GAO's Desirable Characteristics with Respect to Nanotechnology EHS Research	45
Table 5: Elements of Desirable Characteristics of National Strategies	61
Table 6: Collaborative Nanotechnology Environmental, Health, and Safety Research Agreements	66

Figures

Figure 1: The NNI Structure   11

Figure 2: Nanotechnology EHS Research Funding Reported under Program Component Area 7 by NSTC for All NNI Member Agencies, Fiscal Years 2006 through 2010   16

Figure 3: Percentage of Total NNI Research Funding Represented by Each Program Component Area, Fiscal Years 2006 through 2010   17

Figure 4: Categories of Nanomaterials Studied by Seven NNI Member Agencies' EHS Research Projects, Fiscal Year 2010   25

Figure 5: Nanomaterials Studied by Seven NNI Member Agencies' EHS Research Projects, Fiscal Year 2010   28

Figure 6: Usefulness of Formal Collaboration Mechanisms for Nanotechnology EHS Research, According to Questionnaire Respondents   35

Figure 7: Usefulness of Collaborative EHS Research or Related Activities with NNI Member Agencies, According to Questionnaire Respondents   36

Figure 8: Challenges to Collaboration on Nanotechnology EHS Research, According to Questionnaire Respondents   38

Figure 9: Effectiveness of Mechanisms for Obtaining Input on the NNI Strategic Planning for EHS Research, According to Questionnaire Respondents   30

Figure 10: Effectiveness of NNI EHS Research Strategies at Addressing Nanotechnology EHS Research Needs, According to Questionnaire Respondents   40

Figure 11: Frequency with Which Questionnaire Respondents Reported Obtaining Information on the Potential EHS Risks of Nanotechnology from NNI Member Agencies   41

Figure 12: Frequency with Which Questionnaire Respondents Reported Obtaining Information on the Potential EHS Risks of Nanotechnology from Nongovernmental Sources   42

Abbreviations

CPSC	Consumer Product Safety Commission
DOD	Department of Defense
EHS	environmental, health, and safety
EPA	Environmental Protection Agency
FDA	Food and Drug Administration
IANH	International Alliance for NanoEHS Harmonization
ICON	International Council on Nanotechnology
NCI	National Cancer Institute
NEHI	Nanotechnology Environmental and Health Implications
NGO	nongovernmental organizations
NIEHS	National Institute of Environmental Health and Science
NIH	National Institutes of Health
NIOSH	National Institute for Occupational Safety and Health
NIST	National Institute of Standards and Technology
NNCO	National Nanotechnology Coordination Office
NNI	National Nanotechnology Initiative
NRC	National Research Council
NSET	Nanoscale Science, Engineering, and Technology
NSF	National Science Foundation
NSTC	National Science and Technology Council
OMB	Office of Management and Budget
OSTP	Office of Science and Technology Policy
PCA	program component area
PCAST	President's Council of Advisors on Science and Technology
R&D	research and development
Recovery Act	American Recovery and Reinvestment Act of 2009
SAIC	Science Applications International Corporation
UK	United Kingdom

This is a work of the U.S. government and is not subject to copyright protection in the United States. The published product may be reproduced and distributed in its entirety without further permission from GAO. However, because this work may contain copyrighted images or other material, permission from the copyright holder may be necessary if you wish to reproduce this material separately.

May 21, 2012

The Honorable Barbara Boxer
Chairman
Committee on Environment
  and Public Works
United States Senate

Dear Madam Chairman:

Nanotechnology is the understanding and control of matter at approximately 1 and 100 nanometers, known as the nanoscale. For illustration, a sheet of paper is about 100,000 nanometers thick, a human hair is about 80,000 nanometers in diameter, and three gold atoms lying side by side are about 1 nanometer long. Unusual properties can emerge in materials manufactured at the nanoscale—including electrical, magnetic, mechanical, optical, and thermal properties—that differ in important ways from the properties of conventionally scaled materials. Materials at this scale that have been manufactured are sometimes referred to as engineered nanomaterials.[1] Nanotechnology has enabled or facilitated novel research in areas such as computing, medicine, energy conversion and storage, water purification, agriculture and food systems, synthetic biology, aerospace, and geoengineering. Consumer products that incorporate nanotechnology are as diverse as clothing, cosmetics, household appliances, and sporting goods. For example, nanoscale particles of titanium dioxide used in sunscreens act as physical filters that absorb UV light without affecting the opacity of the product. A 2010 study estimated that values for products enabled by nanotechnology were worth about $91 billion in the United States and $254 billion worldwide in 2009.[2] Trends suggest that the number of nanotechnology products and workers employed in related fields will double every 3 years worldwide, achieving a $3 trillion market and 6 million workers by 2020.

---

[1]Nanomaterials may also occur naturally, such as components of volcanic ash or ocean spray, or be created incidentally, including as byproducts of welding. Our review focuses on engineered nanomaterials. When the term nanomaterial is used in our report, it refers to engineered nanomaterials.

[2]Mihail C. Roco, Chad A. Mirkin, and Mark C. Hersam, eds., *Nanotechnology Research Directions for Societal Needs in 2020* (Netherlands: Springer, 2010).

Even with its increasing commercialization, much remains unknown about nanotechnology, including some of the environmental, health, and safety (EHS) impacts of nanomaterials. For example, there are few tools and methods—such as models to predict the behavior of nanomaterials in the environment—for conducting research. It is difficult to assess the risk of nanomaterials because these materials are too varied to generalize how they will behave; nonetheless, risks associated with particular uses of specific nanomaterials can be assessed. Further, there is little information on the number of workers exposed to nanomaterials in the workplace or the effects on human health of such exposure. However, as we reported in 2010, some research indicates that the toxicity of certain nanomaterials, such as some forms of carbon nanotubes and nanoscale titanium dioxide, may pose a risk to human health.[3]

The National Nanotechnology Initiative (NNI) is a federal interagency program that seeks to expedite the discovery, development, and deployment of nanoscale science, engineering, and technology to serve the public good through coordinated research and development aligned with the missions of its member agencies.[4] The NNI informs and influences the federal nanotechnology budget and planning process through its member agencies and through the National Science and Technology Council (NSTC), an entity administered by the Office of Science and Technology Policy (OSTP) that is the principal means by which the executive branch coordinates science and technology policy. The NSTC's Committee on Technology coordinates the NNI under the Nanoscale Science, Engineering, and Technology (NSET) Subcommittee. The National Nanotechnology Coordination Office (NNCO), which reports to the Committee on Technology, provides technical and administrative

---

[3]GAO, *Nanotechnology: Nanomaterials Are Widely Used in Commerce, but EPA Faces Challenges in Regulating Risk,* GAO-10-549 (Washington, D.C.: May 25, 2010).

[4]As of 2010, the NNI member agencies include the Departments of Defense, Education, Energy, Homeland Security, Justice, Labor, State, Treasury, and Transportation; the Department of Agriculture's National Institute of Food and Agriculture and Forest Service; the Department of Commerce's Bureau of Industry and Security, National Institute of Standards and Technology, and the U.S. Patent and Trademark Office; the Department of Health and Human Services' Food and Drug Administration, National Institute for Occupational Safety and Health, and the National Institutes of Health; the Department of the Interior's U.S. Geological Survey; the Consumer Product Safety Commission; Environmental Protection Agency; intelligence agencies; the International Trade Commission; National Aeronautics and Space Administration; Nuclear Regulatory Commission; National Science Foundation; Office of Management and Budget; and Office of Science and Technology Policy.

support to the NSET. The NNI, through the NNCO and appropriate agencies, among others, is directed to (1) establish goals, priorities, and metrics for evaluation for federal nanotechnology research, development, and other activities; (2) invest in federal research and development programs in nanotechnology and related sciences to achieve these goals; and (3) provide for interagency coordination of federal nanotechnology research, development, and other activities.

The federal nanotechnology commitment is significant. Cumulative reported NNI funding since fiscal year 2001 amounts to more than $14 billion. Nanotechnology research projects and activities undertaken by the NNI member agencies are categorized into eight program component areas.[5] Agencies report their research funding for each program component area to the Office of Management and Budget (OMB) as part of the annual federal budget process. The seventh program component area comprises research "primarily directed at understanding the EHS impacts of nanotechnology development and corresponding risk assessment, risk management, and methods for risk mitigation," which we refer to in this report as EHS research.

The NSTC issued three key NNI strategy documents in 2011 that, as a whole, seek to establish shared goals and research needs; describe NNI member agencies' activities; and provide guidance for agency leaders, program managers, and the research community regarding planning and implementation of nanotechnology research and development investments and activities, including those related to EHS research.[6] Jointly, these documents form a national strategy for federal nanotechnology efforts—such strategies tend to cut across sectors of the

---

[5]The eight program component areas are (1) fundamental nanoscale phenomena and processes; (2) nanomaterials; (3) nanoscale devices and systems; (4) instrumentation research, metrology, and standards for nanotechnology; (5) nanomanufacturing; (6) major research facilities and instrumentation acquisition; (7) environment, health, and safety; and (8) education and societal dimensions.

[6]These three strategy documents are: (1) NSTC, Committee on Technology, Subcommittee on Nanoscale Science, Engineering, and Technology, *National Nanotechnology Initiative Strategic Plan* (February 2011); (2) NSTC, Committee on Technology, Subcommittee on Nanoscale Science, Engineering, and Technology, *National Nanotechnology Initiative Environmental, Health, and Safety Research Strategy* (October 2011); and (3) NSTC, Committee on Technology, Subcommittee on Nanoscale Science, Engineering, and Technology, *The National Nanotechnology Initiative Supplement to the President's 2012 Budget* (February 2011).

economy and levels of government and involve a range of public and private organizations and entities.[7] The 2011 NNI strategic plan provides a vision and overarching goals for the program, including responsible development of nanotechnology, which includes EHS research. The 2011 NNI EHS research strategy expands on the goal of the responsible development of nanotechnology by describing the state of science and research needed to ensure that nanotechnology provides maximum benefits to the environment and human well-being. The NNI Supplement to the President's 2012 Budget serves as the annual report for the NNI and describes recent activities by the NNI member agencies and funding of the NNI, including the funding level for the program component area related to EHS research. These NNI strategy documents are the principal public record of the NSTC's coordinating activities and reflect the NSTC's guidance to agencies as they separately determine their budgets and activities for nanotechnology EHS research.

In prior work, we have identified six desirable characteristics for national strategies and reviewed strategies that the federal government has used to plan for and report on crosscutting issues including emergency preparedness, financial literacy, influenza, and terrorism for these characteristics.[8] These desirable characteristics help shape the policies, programs, priorities, resource allocations, and standards that would enable federal agencies and other stakeholders to implement the strategies and achieve the identified results. National strategies that address these characteristics offer policymakers and implementing agencies a management tool that can help ensure accountability and more effective results.

---

[7]These documents also describe mechanisms for outreach and collaboration with stakeholders who are involved in nanotechnology EHS research, including those working in state and local governments, academia, companies, and nongovernmental organizations.

[8]GAO, *National Capital Region: 2010 Strategic Plan is Generally Consistent with Characteristics of Effective Strategies,* GAO-12-276T (Washington, D.C.: Dec. 7, 2011); *Influenza Pandemic: Further Efforts Are Needed to Ensure Clearer Federal Leadership Roles and an Effective National Strategy,* GAO-07-781 (Washington, D.C.: Aug. 14, 2007); *Financial Literacy and Education Commission: Further Progress Needed to Ensure an Effective National Strategy,* GAO-07-100 (Washington, D.C.: Dec. 4. 2006); *Combating Terrorism: Evaluation of Selected Characteristics in National Strategies Related to Terrorism,* GAO-04-408T (Washington, D.C.: Feb. 3, 2004).

In this context, you asked us to review federal nanotechnology EHS research. This report examines (1) changes in federal funding for nanotechnology EHS research from fiscal years 2006 to 2010; (2) the nanomaterials that NNI member agencies focused on in their EHS research in fiscal year 2010; (3) the extent to which NNI member agencies collaborate with stakeholders on nanotechnology EHS research and related strategies; and (4) the extent to which NNI strategy documents address desirable characteristics of national strategies.

To conduct this work, we reviewed EHS research efforts funded by seven NNI agencies, which collectively funded 93 percent of EHS research dollars in fiscal year 2010: the National Science Foundation (NSF), National Institutes of Health (NIH), Environmental Protection Agency (EPA), National Institute for Occupational Safety and Health (NIOSH), Food and Drug Administration (FDA), National Institute of Standards and Technology (NIST), and the Consumer Product Safety Commission (CPSC). The first six of these agencies represent the top six providers of EHS research funding from fiscal year 2006 to fiscal year 2010, and CPSC has an important role in ensuring the safe use of nanotechnology in consumer products.

To examine recent changes in federal funding for nanotechnology EHS research, we reviewed information published by NSTC in the NNI Supplements to the President's Budget. Specifically, we reviewed funding reported in each program component area for fiscal years 2006 through 2010 for all NNI member agencies funding nanotechnology research.[9] For the dollar amounts that we adjusted for inflation, we used the Biomedical Research and Development Price Index to report funding in constant 2010 dollars.[10] We consulted with agency and OMB officials to determine

---

[9]See NSTC, Committee on Technology, Subcommittee on Nanoscale Science, Engineering, and Technology, *The National Nanotechnology Initiative Supplement to the President's 2012 Budget* (February 2011); *The National Nanotechnology Initiative Supplement to the President's 2011 Budget* (February 2010); *The National Nanotechnology Initiative Supplement to the President's 2010 Budget* (May 2009); *The National Nanotechnology Initiative Supplement to the President's 2009 Budget* (September 2008); and *The National Nanotechnology Initiative Supplement to the President's 2008 Budget* (July 2007).

[10]The Biomedical Research and Development Price Index is prepared by the Department of Commerce, Bureau of Economic Analysis for use of and publication by NIH. Information about the index is available at http://officeofbudget.od.nih.gov/pdfs/FY13/BRDPI_Proj_Jan_2012_Final.pdf.

the type of budget information reported in these documents for the agencies' investments. For fiscal year 2010, the most recent year for which agencies' nanotechnology investment data were available, we also collected quantitative and qualitative project-level data on all research projects that the seven selected agencies had categorized as EHS research (program component area 7). The agencies do not report project-level data to OMB annually but report their total EHS research funding to OMB annually for inclusion in the NNI budget supplements published by NSTC. Therefore, we reviewed data on the individual projects included in the agencies' total EHS research funding in 2010. To assess the reliability of the agencies' data, we obtained information regarding the data and the information systems used to produce and store them. We also reviewed related supporting documentation and consulted with agency officials. From our assessment, we determined that the data were sufficiently reliable for our purposes. We analyzed these data and consulted with agency officials to assess whether the agencies had appropriately categorized as EHS research the projects they reported as program component area 7—research "primarily directed at understanding the EHS impacts of nanotechnology development and corresponding risk assessment, risk management, and methods for risk mitigation." We also reviewed our 2008 report in this area,[11] which found that 22 of 119 fiscal year 2006 projects reported as EHS research were miscategorized, and recommended that the Director of OSTP, in consultation with the Directors of NNCO and OMB, provide better guidance to agencies on how to report nanotechnology EHS research. To identify the nanomaterials studied by the fiscal year 2010 EHS research projects, we reviewed qualitative project data such as abstracts for the projects we determined were primarily directed at EHS and consulted with agency officials. We then grouped the nanomaterials into five categories.[12]

To determine the extent to which the NNI agencies collaborate with stakeholders on nanotechnology EHS research and related strategies, we (1) discussed with agency officials how their agencies collaborate on

---

[11]GAO, *Nanotechnology: Better Guidance Is Needed to Ensure Accurate Reporting of Federal Research Focused on Environmental, Health, and Safety Risks*, GAO-08-402 (Washington, D.C.: Mar. 31, 2008).

[12]The five categories are carbon-based nanomaterials, metal-based nanomaterials, semiconductor-based nanomaterials, organic nanomaterials, and other nanomaterials.

nanotechnology EHS research and NNI's role in facilitating that collaboration and (2) obtained documentation on these collaborative efforts. We conducted a review of formal collaborative efforts that focused on nanotechnology EHS research initiated from February 2008 to October 2011. We administered a web-based questionnaire to a nonprobability sample of 223 nonfederal stakeholders, including those affiliated with academia, companies, nongovernmental organizations (NGO), and state and local governments, to obtain their views on collaboration with the NNI member agencies on EHS research and the NNI EHS research strategies. Of the 223, 138 completed the questionnaire, for an overall response rate of 62 percent. This sample included individuals who had expertise in the field of nanotechnology, had interacted with the NNI in the past few years, or who were representatives of organizations and companies suggested to us through our scoping interviews.

To determine the extent to which the NNI strategy documents address desirable characteristics of national strategies, we analyzed three key NNI strategy documents related to nanotechnology EHS for the presence of six desirable characteristics we previously identified for national strategies. We also reviewed NNI member agencies' management documents for the use of performance information related to nanotechnology EHS research and the extent to which they were linked with the NNI national priorities. Appendix I presents a more detailed description of our scope and methodology.

We conducted this performance audit from February 2011 through May 2012 in accordance with generally accepted government auditing standards. Those standards require that we plan and perform the audit to obtain sufficient, appropriate evidence to provide a reasonable basis for our findings and conclusions based on our audit objectives. We believe that the evidence obtained provides a reasonable basis for our findings and conclusions based on our audit objectives.

Background

Nanomaterials come in a variety of forms based both on their chemical composition and physical structure. For example, carbon-based nanomaterials can be produced in a number of physical structures such as sheets (graphene), tubes (carbon nanotubes), and particles (carbon black). Nanomaterials can enter the marketplace as materials themselves, as intermediates that either have nanoscale features or incorporate nanomaterials, and as final nano-enabled products.

The use of nanomaterials in commercial applications raises questions about the potential risks that might arise from exposures to nanomaterials and the differences in exposure during their manufacture, use, and disposal. For example, some, but not all, research studies have shown adverse respiratory or cellular effects in animals exposed to some types of carbon nanotubes. Observed effects include early onset and persistence of pulmonary fibrosis and interference with cell division.[13] The risk posed by a material is a combination of the hazard or negative effect that material may have on an organism and the extent of the organism's exposure to that material. Therefore, a highly poisonous material—that is, one with high hazard—may nonetheless pose little risk if susceptible groups have little or no contact with the material. For instance, many household chemicals are hazardous to human health but pose little risk when exposure is limited by safe handling. Conversely, a material with relatively mild health effects may pose a large risk if people or the environment are exposed to large amounts or over prolonged periods.

Efforts to examine EHS risks are complicated by the variety of nanomaterials available for research and commerce. According to the National Research Council (NRC) of the National Academies, a nonprofit institution providing scientific advice, there are currently gaps in understanding what factors or underlying phenomena contribute to potential hazards from nanomaterials.[14] In 2010, we reported that size alone is not a sufficient indicator of the potential risk of a material.[15] Research on carbon nanotubes, for example, suggests that length, purity, and other factors may be involved as well. Similarly, data on potential exposures of people and the environment are incomplete. In the absence of clear indicators of risk, some organizations sponsoring research on nanomaterials have identified specific materials or classes of materials as high-priority targets for EHS testing based on availability and commercial interest. For example, as part of a testing program of the EHS risks of nanomaterials, the Organisation for Economic Co-operation and

---

[13]In this report, we did not attempt to summarize the current scientific literature on the potential hazards of or exposure to nanomaterials. We most recently reviewed what is known about the potential human health and environmental risks of nanomaterials in 2010, see GAO-10-549.

[14]NRC, *A Research Strategy for Environmental, Health, and Safety Aspects of Engineered Nanomaterials* (2012).

[15]GAO-10-549.

Development has developed a list of 13 representative manufactured nanomaterials now or soon to be in commerce as priority testing targets.[16,17]

Regulatory agencies, like FDA, CPSC, and EPA, may not have complete information on the uses or risks of some nanomaterials. In 2010, we reported that uncertainties persist about how to evaluate the safety of engineered nanomaterials in food and that nanomaterials may enter the food supply in certain products generally recognized as safe without FDA's knowledge.[18] We also reported in 2010 that EPA has taken a variety of actions to better understand and regulate the risks of nanomaterials but that the agency faces challenges that might impede its ability to regulate nanomaterials effectively.[19] For example, we reported that the Toxic Substances Control Act gives EPA authority to issue rules requiring companies to submit certain information about chemicals. EPA recently amended a regulation to require companies to report certain information regarding production of chemicals above certain thresholds but those thresholds may not capture nanomaterials if they are produced in amounts below the thresholds. EPA currently has no plans to further reduce the thresholds. In addition, according to officials at CPSC, which is charged with protecting the public from unreasonable risks of injury or death from thousands of types of consumer products, CPSC does not have the statutory authority to require pre-market approval for products, including those incorporating nanomaterials.

---

[16]The Organisation for Economic Co-operation and Development is an organization of 34 national governments, operating by consensus, that fosters dialogue among members to discuss, develop, and refine economic and social policies and provides an arena for setting rules when multilateral agreements are necessary.

[17]The nanomaterials the Organisation for Economic Co-operation and Development selected are fullerenes, single-walled carbon nanotubes, multi-walled carbon nanotubes, silver nanoparticles, iron nanoparticles, titanium dioxide, aluminum oxide, cerium oxide, zinc oxide, silicon dioxide, dendrimers, nanoclays, and gold nanoparticles.

[18]GAO, *Food Safety: FDA Should Strengthen Its Oversight of Food Ingredients Determined to Be Generally Recognized as Safe (GRAS)*, GAO-10-246 (Washington, D.C.: Feb. 3, 2010). Subsequent to this report, FDA issued draft guidance on April 20, 2012, on assessing the effects of significant manufacturing changes, including any involving nanotechnology, on the identity, safety, or regulatory status of food substances.

[19]GAO-10-549.

The NNI was codified in law by the 21st Century Nanotechnology Research and Development Act in 2003.[20] The act requires the President to implement a national nanotechnology program and charges the NSTC itself—or through a subgroup—with overseeing the planning, management, and coordination of the program. The NSTC carries out these tasks through the NSET Subcommittee, which includes a co-chair from OSTP as well as representatives from the member agencies of the NNI. The NSET oversees working groups, including the Nanotechnology Environmental and Health Implications (NEHI) working group which supports federal activities to protect public health and the environment. Figure 1 provides an overview of the organizational structure of the NNI.

---

[20]Pub. L. No. 108-153 (2003).

Figure 1: The NNI Structure

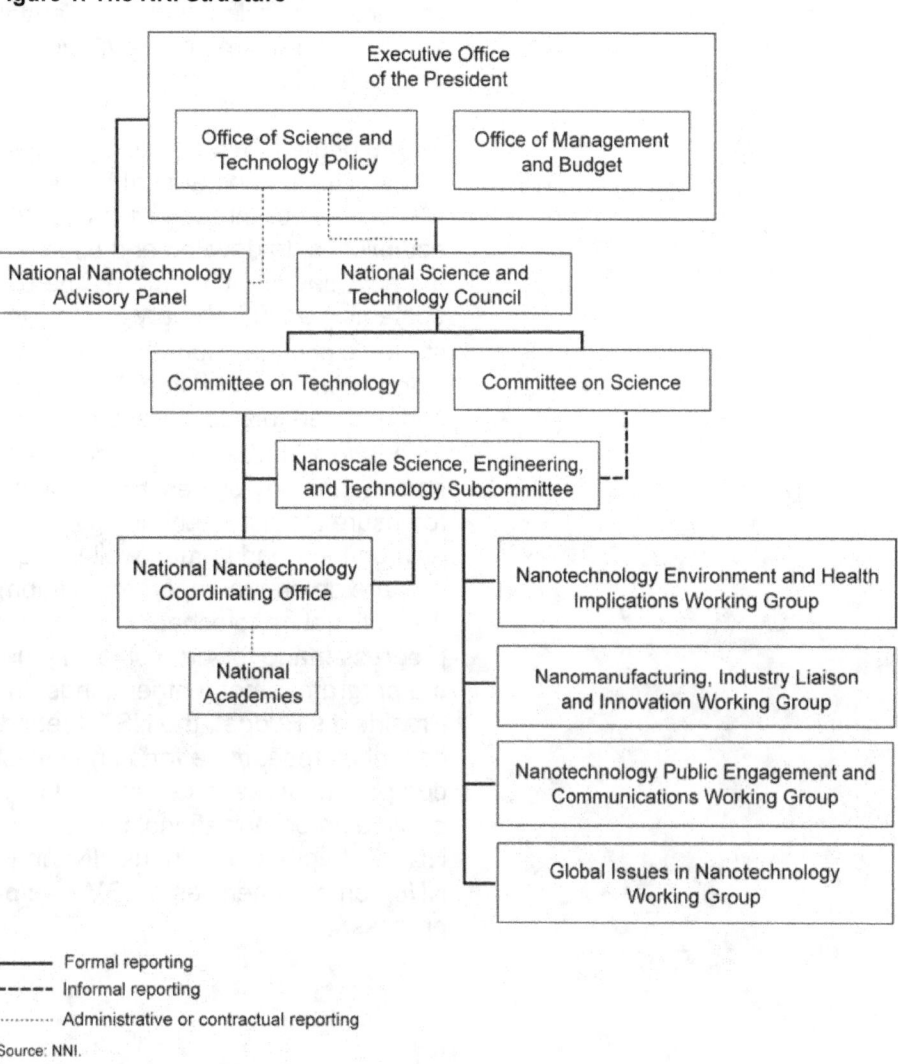

Formal reporting
----- Informal reporting
.... Administrative or contractual reporting

Source: NNI.

The NNI does not fund research directly; rather, each of its member agencies determines its nanotechnology activities based on its individual mission and priorities. The NNI provides a framework for a comprehensive nanotechnology research and development program by establishing shared goals, priorities, and strategies among member agencies; and providing avenues for member agencies to leverage the resources of all participating agencies. The four goals of the NNI are to (1) advance a world-class nanotechnology research and development program; (2) foster the transfer of new technologies into products for commercial and public benefit; (3) develop and sustain educational resources, a skilled workforce, and the supporting infrastructure and tools to advance nanotechnology; and (4) support responsible development of nanotechnology.

Efforts of the NNI member agencies are reported through triennial strategic plans and annual budget supplements. The act directs the NSTC, itself or through an appropriate subgroup it designates or establishes, to develop and update every 3 years a strategic plan to guide the activities of the program. The NSTC published its most recent strategic plan in February 2011.[21] In addition to the statutorily required strategic plan, in 2008, the NSTC published a *Strategy for Nanotechnology-Related Environmental, Health, and Safety Research* and updated this document in 2011.[22] The EHS strategies published in 2008 and 2011 expand on the goal of the responsible development of nanotechnology by describing the state of science and research needed to ensure that nanotechnology provides maximum benefits to the environment and human well-being. The act also requires the NSTC to prepare an annual report to be submitted to congressional committees on the national nanotechnology program's budget and an analysis of the progress made toward achieving the goals and priorities established for the program, among other things. In the NNI Supplements to the President's Budget, the NSTC reports overall NNI spending and describes research efforts and investments within eight program component areas, as seen in table 1. These program component areas provide an organizational framework for categorizing the activities of the NNI. Data presented in the NNI annual supplements is collected from the NNI member agencies by OMB as part of the annual budget formulation process.

---

[21]NSTC, Committee on Technology, Subcommittee on Nanoscale Science, Engineering, and Technology, *National Nanotechnology Initiative Strategic Plan* (February 2011).

[22]NSTC, Committee on Technology, Subcommittee on Nanoscale Science, Engineering, and Technology, *National Nanotechnology Initiative Strategy for Nanotechnology-Related Environmental, Health, and Safety Research* (February 2008) and *National Nanotechnology Initiative Environmental, Health, and Safety Research Strategy* (October 2011).

Table 1: Program Component Areas

No.	Title	Description
1	Fundamental Nanoscale Phenomena and Processes	Discovery and development of fundamental knowledge pertaining to new phenomena in the physical, biological, and engineering sciences that occur at the nanoscale. Elucidation of scientific and engineering principles related to nanoscale structures, processes, and mechanisms.
2	Nanomaterials	Research aimed at the discovery of novel nanoscale and nanostructured materials and at a comprehensive understanding of the properties of nanomaterials (ranging across length scales, and including interface interactions). Research and development (R&D) leading to the ability to design and synthesize, in a controlled manner, nanostructured materials with targeted properties.
3	Nanoscale Devices and Systems	R&D that applies the principles of nanoscale science and engineering to create novel, or to improve existing, devices and systems. Includes the incorporation of nanoscale or nanostructured materials to achieve improved performance or new functionality. To meet this definition, the enabling science and technology must be at the nanoscale, but the systems and devices themselves are not restricted to that size.
4	Instrumentation Research, Metrology, and Standards for Nanotechnology	R&D pertaining to the tools needed to advance nanotechnology research and commercialization, including next-generation instrumentation for characterization, measurement, synthesis, and design of materials, structures, devices, and systems. Also includes R&D and other activities related to development of standards, including standards for nomenclature, materials characterization and testing, and manufacture.
5	Nanomanufacturing	R&D aimed at enabling scaled-up, reliable, and cost-effective manufacturing of nanoscale materials, structures, devices, and systems. Includes R&D and integration of ultra-miniaturized top-down processes and increasingly complex bottom-up or self-assembly processes.
6	Major Research Facilities and Instrumentation Acquisition	Establishment of user facilities, acquisition of major instrumentation, and other activities that develop, support, or enhance the nation's scientific infrastructure for the conduct of nanoscale science, engineering, and technology R&D. Includes ongoing operation of user facilities and networks.
7	Environment, Health, and Safety	Research primarily directed at understanding the environmental, health, and safety impacts of nanotechnology development and corresponding risk assessment, risk management, and methods for risk mitigation.
8	Education and Societal Dimensions	Education-related activities such as development of materials for schools, undergraduate programs, technical training, and public communication, including outreach and engagement. Research directed at identifying and quantifying the broad implications of nanotechnology for society, including social, economic, workforce, educational, ethical, and legal implications.

Source: NSTC, 2011 NNI Strategic Plan.

The act requires triennial external reviews of the national nanotechnology program. Specifically, the act requires the NNCO to contract with the NRC to conduct the triennial evaluations of the national nanotechnology program. The NRC draws on expertise from outside government, including from academia, companies, and NGOs. The NRC completed reviews in 2002, 2006, and 2009 based on the work of 15 to 23 panelists chosen by the NRC and identified in the review. In addition, the NRC also published an independent research strategy to address EHS aspects of nanomaterials in 2012.[23]

The act also requires the President to establish or designate a National Nanotechnology Advisory Panel. The advisory panel, by statute, must consist primarily of members from academic institutions and industry, and panel members must be qualified to provide advice and information on nanotechnology research, development, demonstrations, education, technology transfer, commercial application, or societal and ethical concerns. Since 2004, the President has designated the President's Council of Advisors on Science and Technology (PCAST) to function as the advisory panel. PCAST members are generally senior managers in major corporations and academia selected for diverse expertise in science and technology issues. The advisory panel is required to report not less frequently than once every 2 fiscal years on its assessment of the national nanotechnology program and recommendations for ways to improve the program. The advisory panel has produced three assessments to date, in 2005, 2008, and 2010. The first and second assessments were authored by a subset of PCAST membership, and the third assessment was authored by a working group of three PCAST members and additional external experts, but PCAST as a whole approved the assessments. The first and second assessments created Nanotechnology Technical Advisory Groups, which were comprised of approximately 10 members who provided written responses to questionnaires developed by PCAST. The membership of those groups is not identified in the assessments. The external members of the third assessment's working group were selected by PCAST members and are identified in PCAST's assessment.[24]

---

[23]NRC, *A Research Strategy for Environmental, Health, and Safety Aspects of Engineered Nanomaterials* (2012). We did not fully evaluate this strategy in the context of federal EHS research efforts because it was issued in draft form in January 2012.

[24]The working group invited additional experts to speak at public meetings, including individuals working in academia and the corporate sector.

Funding of EHS Research by NNI Member Agencies Has Increased, but Quality of Funding Data Is Uncertain

NSTC reported more than a doubling of funding for nanotechnology EHS research by NNI member agencies from fiscal years 2006 to 2010.[25] Reported EHS funding also rose as a percentage of total NNI funding during this period, ending up at about 5 percent in 2010. We also identified several reporting problems related to the continued absence of detailed guidance on how agencies should report funding for their nanotechnology research, raising concerns about the quality of EHS funding data reported.

EHS Research Funding by NNI Member Agencies More Than Doubled From 2006 to 2010 and Amounted to About 5 Percent of Total NNI Funding in 2010

From fiscal years 2006 to 2010, NSTC reported more than a doubling of NNI member agencies' funding for nanotechnology EHS research in the NNI Supplements to the President's Budget—from approximately $38 million[26] to $90 million.[27] As shown in figure 2, most of this funding was reported to be by NSF, NIH, and EPA, and the largest increases in reported EHS research funding over this period were at NIH and EPA.

---

[25]This includes funding by the seven agencies we selected (CPSC, EPA, FDA, NIH, NIOSH, NIST, and NSF) as well as by the other NNI members funding nanotechnology EHS research (the Departments of Defense, Energy, and Agriculture).

[26]Adjusted for inflation, the amount of nanotechnology EHS research funding NSTC reported for fiscal year 2006 was approximately $43 million, in 2010 dollars.

[27]This includes funding by the seven agencies we selected as well as by the other NNI members funding nanotechnology EHS research.

Figure 2: Nanotechnology EHS Research Funding Reported under Program Component Area 7 by NSTC for All NNI Member Agencies, Fiscal Years 2006 through 2010

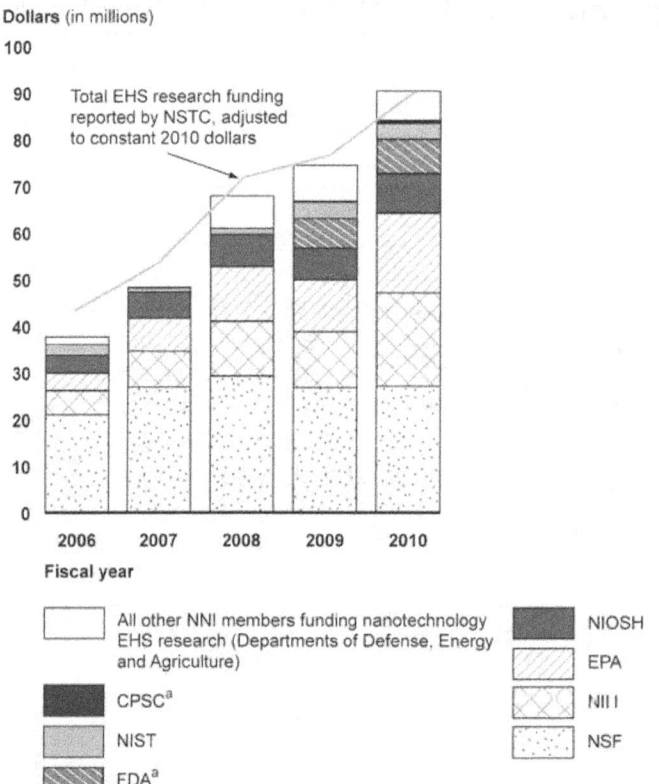

Source: GAO analysis of data reported by NSTC in the annual NNI Supplements to the President's Budget.

Note: This figure does not include research NSTC separately reported as funded by the American Recovery and Reinvestment Act of 2009 (Recovery Act) (Pub. L. No. 111-5 (2009)). The Recovery Act was enacted to, among other things, preserve and create jobs and promote economic recovery. For fiscal year 2009, NSTC reported additional funding for EHS research funded by the Recovery Act, as follows: NSF, $3.4 million; NIH, $8.4 million; and Department of Energy, $0.2 million. For fiscal year 2010, NSTC did not separately report nanotechnology funding from Recovery Act appropriations.

[a]The NNI Supplements to the President's Budget did not include nanotechnology funding data for CPSC and FDA prior to fiscal year 2009.

Figure 3 shows the percentage of NNI funding represented by each program component area (PCA) for all NNI member agencies funding nanotechnology research. Most of the funding reported was for PCA 1 (fundamental nanoscale phenomena and processes), PCA 2 (nanomaterials), and PCA 3 (nanoscale devices and systems). Reported EHS funding rose as a percentage of total NNI funding from fiscal years

2006 to 2010, ending up at about 5 percent in 2010. Specifically, for fiscal year 2010, funding for EHS research was $90 million, while total 2010 nanotechnology funding was $1.9 billion, as reported in the NNI Supplement to the President's 2012 Budget.

Figure 3: Percentage of Total NNI Research Funding Represented by Each Program Component Area, Fiscal Years 2006 through 2010

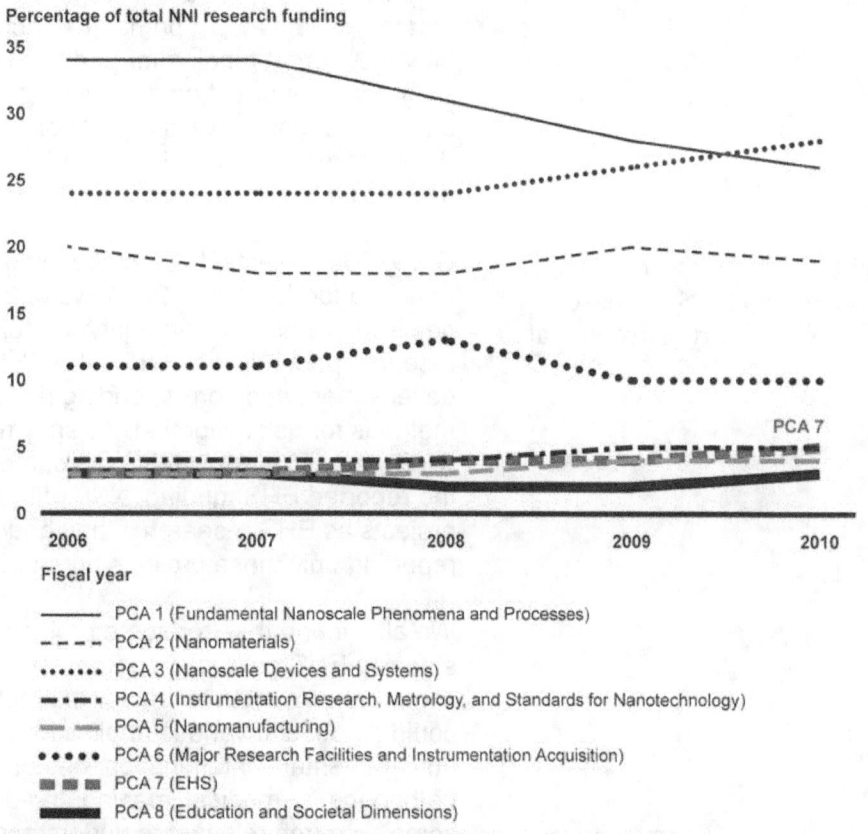

Percentage of total NNI research funding

Fiscal year

——— PCA 1 (Fundamental Nanoscale Phenomena and Processes)
– – – – PCA 2 (Nanomaterials)
•••••• PCA 3 (Nanoscale Devices and Systems)
—•—•— PCA 4 (Instrumentation Research, Metrology, and Standards for Nanotechnology)
— — PCA 5 (Nanomanufacturing)
••••• PCA 6 (Major Research Facilities and Instrumentation Acquisition)
▆▆ ▆▆ ▆▆ PCA 7 (EHS)
▆▆▆▆ PCA 8 (Education and Societal Dimensions)

Source: GAO analysis of data reported by NSTC in the annual NNI Supplements to the President's Budget.

Notes: This figure includes funding by the seven agencies we selected (CPSC, EPA, FDA, NIH, NIOSH, NIST, and NSF) as well as by the other NNI member agencies funding nanotechnology research (the National Aeronautics and Space Administration and the Departments of Defense, Energy, Agriculture, Justice, Homeland Security, and Transportation).

This figure does not include research NSTC separately reported as funded by the Recovery Act. For fiscal year 2009, NSTC reported an additional $511 million for activities funded by the Recovery Act, as follows: PCA 1, $131 million; PCA 2, $178 million; PCA 3, $68 million; PCA 4, $12 million; PCA 5, $29 million; PCA 6, $73 million; PCA 7, $12 million; and PCA 8, $9 million. For fiscal year 2010, NSTC did not separately report agencies' funding from the Recovery Act appropriations.

Reporting Problems Raise Concerns about Quality of EHS Research Funding Data

Based on our review of the NNI member agencies' EHS research funding data published by NSTC in the NNI Supplements to the President's Budget and detailed project data we obtained from the seven selected agencies for fiscal year 2010, we identified several reporting problems that raise concerns about the accuracy, consistency, and completeness of funding data reported and make it difficult to assess changes in federal funding for nanotechnology EHS research over time. We found that (1) for 18 percent of 2010 projects reported as EHS research, it was not clear that the projects were primarily directed at EHS risks; (2) agencies vary in how they report funding for portions of projects that address multiple program component areas or include non-nanotechnology research; (3) agencies did not always report comparable budget data; and (4) NNI does not report updated data after publication of the NNI budget supplements, and the reported data may not reflect all EHS funding.

Eighteen Percent of Projects Reported as EHS Research Were Not Clearly Directed at EHS Risks

Of the 236 projects that the seven agencies reported to us as EHS research for fiscal year 2010, we determined that, for 43 projects (18 percent), it was not clear that the projects met the definition for PCA 7—research primarily directed at the EHS impacts of nanotechnology development and corresponding risk assessment, risk management, and methods for risk mitigation. As shown in table 2, these projects, 20 of which were funded by NSF, accounted for approximately $15 million of the reported EHS funding. NSF officials told us that the agency reported projects as EHS research if they had some relevance to EHS instead of reporting only those projects primarily directed at EHS.

We also found that, for some projects reported by other agencies, studying EHS risks was not the primary purpose of the projects. Instead, some projects focused on, for example, exploring how nanotechnology could be used in various applications, such as remediation of environmental contamination, detection of chemical hazards or pathogens, or medical imaging. In addition, some other projects included some research relevant to understanding EHS impacts of nanotechnology, but it was not clear that the study of EHS issues was the primary purpose. For example, some projects studied various nanotechnology-enabled drugs, but it was not clear to what extent the research was directed at the safety of the drugs versus their efficacy. Other projects were directed at instrumentation and nanotechnology measurement issues that had some relevance for EHS research, but it was not clear whether these should be reported as PCA 7 (EHS research) or PCA 4 (instrumentation research, metrology, and standards for nanotechnology).

Table 2: GAO Analysis of Nanotechnology Projects Reported by Selected Agencies as EHS Research, Fiscal Year 2010

| Agency | Projects seven agencies reported to GAO as EHS research | | GAO analysis of agencies' projects | | | |
| | | | Primarily directed at EHS | | Not clear that projects were primarily directed at EHS | |
	Number[a]	Funding (dollars in millions)	Number	Funding (dollars in millions)	Number	Funding (dollars in millions)
CPSC	8	$0.5	8	$0.5	0	$0
EPA[b]	32	17.6[c]	29	16.9	3	0.8
FDA	29[d]	5.3[c]	20	3.8	9	1.5
NIH	58	27.7[e]	55	26.4[f]	3	1.3[g]
NIOSH	49	9.8[h]	47	9.7	2	0.1
NIST	11	3.4	5	1.6	6	1.8
NSF	49	27.1	29	17.1	20	10.0
Total[i]	236	$91.4[j]	193	$76.0	43	$15.5

Source: GAO analysis of obligations data reported to us by seven agencies.

Notes: The data shown in this table may differ from that reported by NSTC in the NNI Supplement to the President's 2012 Budget, as described below.

[a]For some projects at NIH, NIOSH, and NIST, we grouped together some of the agencies' data entries and counted them together for the purposes of our analyses. Consequently, the numbers of projects reported here may not match the number reported by these agencies elsewhere. Appendix I presents more details on how we grouped projects at these agencies.

[b]According to an EPA official, the total EHS research funding reported to us for fiscal year 2010 represents actual obligations, but the amounts reported for individual projects are estimates.

[c]The total for this agency differs from that reported by NSTC in the NNI Supplement to the President's 2012 Budget because the agency reported its budget authority to OMB for the budget supplement and reported obligations to us.

[d]FDA reported a total of 30 EHS research projects to us; however, for one of those projects, FDA reported that no funds were obligated in fiscal year 2010. Therefore, we excluded that project from our analysis.

[e]The total for this agency differs from that reported by NSTC in the NNI Supplement to the President's 2012 Budget because NIH included in the data reported to us its fiscal year 2010 obligations of funds appropriated by the Recovery Act, which it did not report to OMB for the supplement.

[f]We added $35,044 to NIH's reported total because we identified additional funding for a subproject that we determined should have been included. NIH officials told us that the sub-project was not included because the project abstract and other associated text did not contain any of the key words for the computerized search of projects that NIH uses to develop its list of EHS research.

[g]For two additional NIH projects, we found that the amount of funding for research primarily directed at EHS was overstated. For each of these projects, NIH included the funding for one or more subprojects that were not focused on nanotechnology. We included the overstated amounts in the dollar value for projects where it was not clear that the research was primarily directed at EHS.

[h]The total for this agency differs from that reported by NSTC in the NNI Supplement to the President's 2012 Budget because NIOSH updated its data after initially reporting it to OMB.

[i]Totals may not add because of rounding.

[j]The total dollar value of $91.4 million reported to us differs from the total of $90.2 million reported by NSTC in the NNI Supplement to the President's 2012 Budget for two reasons. First, this table includes only the seven agencies we reviewed, not all agencies that funded EHS research in 2010. Second, four of the seven agencies we reviewed reported data to us that differed from those reported by NSTC.

As previously noted, we recommended in our 2008 report that the Director of OSTP, in consultation with the Directors of NNCO and OMB, provide better guidance to agencies on how to report nanotechnology EHS research.[28] OSTP generally concurred with the report's findings and agreed to review the manner in which agencies respond to the current guidance. However, we found that as of February 2012, updated guidance had not been issued, and the definition of each program component area as documented in OMB Circular A-11 remained the only written guidance available to the agencies for reporting their nanotechnology EHS research.[29] OMB Circular A-11 does not address what types of projects should or should not be categorized as EHS research, such as how to determine if a project is "primarily" directed at EHS impacts or how to report projects focused on environmental or health-related applications of nanotechnology. For example, with respect to drugs, HHS commented that the Federal Food, Drug, and Cosmetic Act requires that both safety and effectiveness be evaluated, in relation to one another. As HHS noted, both traits are relevant to the overall health impact of a drug product. However, OMB Circular A-11 does not address whether funding for research on the effectiveness of nanotechnology-enabled drugs should be considered nanotechnology EHS research under the definition of PCA 7. It also does not address how to report EHS-related projects that are focused on instrumentation and metrology, and we found that agencies have used different criteria to determine the program component area under which to report such projects as well.

NNCO and FDA officials told us that the reporting of EHS research was a regular agenda item at meetings of the NEHI working group during development of the 2011 NNI EHS research strategy. However, the strategy used information reported via a different OMB data call, one that NNCO and OMB officials said was much more detailed and cast a broader net than the annual nanotechnology data calls OMB conducts for the annual NNI budget supplements (the "crosscut"). The 21st Century Nanotechnology Research and Development Act requires annual reporting of spending for each program component area, and the crosscut collects EHS research funding data based on the definition of PCA 7—research primarily directed at the EHS impacts of nanotechnology

---

[28]GAO-08-402.

[29]OMB Circular No. A-11, Preparation, Submission, and Execution of the Budget, pt. 2, § 84 (July 2010).

development and corresponding risk assessment, risk management, and methods for risk mitigation. However, the data call for the 2011 research strategy was, according to OMB's instructions to agencies, "not limited to research and development efforts whose primary purpose is to understand and address potential risks to health and to the environment posed by this technology—as is called for in the overall crosscut for nanotechnology." Consequently, discussions of how to report projects for use in developing the 2011 research strategy do not necessarily provide guidance to agencies on how to determine whether their research meets the definition of PCA 7, or how to apportion funding for projects that address multiple program component areas.

Agencies Vary in How They Report Funding for Portions of Projects

In our 2008 report, we found that neither NSET nor OMB had provided guidance on whether or how to apportion funding for a single research project to more than one program component area, if appropriate.[30] In this review, we found that agencies have different approaches to reporting funding for projects that address multiple program component areas or include non-nanotechnology research, and the guidance available to the agencies—the PCA definitions in OMB Circular A-11—still does not address this. We found that, for fiscal year 2010 projects where only a portion of the project was focused on nanotechnology EHS research, NIST and NSF assigned only a portion of the total funding to PCA 7. At NIH, grants or contracts may be either single projects or multiproject "parent" grants or contracts, which have multiple individual subprojects. According to NIH officials, NIH policy does not allow the funding for a single project or subproject to be subdivided across program component areas, and NNI policy requires that the individual totals of the program component areas sum to the agency's total nanotechnology funding. As a consequence of these two policies, if a grant or contract is a single project, then NIH reports the entire value of the individual grant or contract in only one program component area. NIH officials told us that if a grant or contract has subprojects, NIH reports the funding for an individual subproject in only one program component area, but the funding for the different subprojects can be reported in different program component areas, as applicable.

At NIOSH, agency officials told us that, for projects that include non-nanotechnology research, they track and report the nanotechnology

---

[30]GAO-08-402.

research portions separately and that the agency reports all of its nanotechnology research as EHS because NIOSH's mission aligns with PCA 7. The NIOSH mission includes conducting research on the potential implications of nanotechnology on worker health and safety, as well as conducting research on potential applications of nanotechnology to solve worker health and safety problems. FDA officials also told us that they report all of their regulatory science research regarding nanotechnology as EHS because it is conducted to support the agency's mission, which they said aligns with PCA 7. As described in FDA's *Strategic Priorities* document, the agency's mission includes ensuring both the safety and the effectiveness of drugs, biological products, and medical devices, as well as the safety of foods and cosmetics.[31]

Agencies Did Not Always Report Comparable Budget Data

Data on nanotechnology funding published in the annual NNI Supplements to the President's Budget are not comparable over time or across agencies because the data do not always represent the same type of budget information. The NNI budget supplements report funding as "actual agency investments" but do not specify what type of data are reported as actual agency investments, and until recently, OMB's instructions to agencies for providing these data did not specify what should be reported.[32] Officials from OMB told us that agencies report budget authority for their actual agency investments,[33] but officials from the seven agencies we reviewed generally told us that they had reported obligations as their actual agency investments for fiscal years 2006 to

---

[31]FDA, *Strategic Priorities 2011-2015*.

[32]OMB's data calls for 2007 to 2009 actual agency investments did not specify what type of budget data should be reported; the data call for 2010 data asked agencies to report budget authority for their actual agency investments.

[33]Budget authority is authority provided by federal law to enter into financial obligations that will result in immediate or future outlays involving federal government funds. Budget authority includes (1) appropriations, (2) borrowing authority, (3) contract authority, and (4) authority to obligate and expend offsetting receipts and collections. Appropriations represent budget authority to incur obligations and make payments from the Treasury for specified purposes. Appropriations do not represent cash actually set aside in the Treasury for purposes specified in the appropriation act; they represent amounts that agencies may obligate during the period of time specified in the respective appropriation acts. See GAO, *A Glossary of Terms Used in the Federal Budget Process*, GAO-05-734SP (Washington, D.C.: Sept. 1, 2005).

2010.[34] However, an agency's budget authority and obligations are not always the same amounts in a given year. Furthermore, agencies may not always report the same type of budget information from year to year. For example, for fiscal year 2009, FDA reported obligations, but then reported budget authority for fiscal year 2010 because obligations data were not available, according to FDA officials.

NNI Does Not Report Updated Data after Publication of Budget Supplements, and Reported Data May Not Reflect All EHS Funding

Funding data published by NSTC in the NNI Supplements to the President's Budget are not always final because agencies sometimes make changes to their data after they are published as actual agency investments in the budget supplements. Consequently, outdated data can remain available in NNI documents and on the NNI website, which publishes nanotechnology research funding data from the budget supplements in response to OMB's 2009 Open Government Directive.[35] For example, NIOSH officials told us that their fiscal year 2010 actual agency investment in EHS research reported in the NNI Supplement to the President's 2012 Budget was an estimate because the timing of OMB's call for the data was earlier than usual. According to NIOSH officials, a second data call occurred later in the fiscal year, at which time the agency was able to report an accurate actual funding level. However, as of February 2012, NIOSH's estimated data have not been updated on the NNI website. In addition, we found that when NIST reported its 2009 nanotechnology research funding from the American Recovery and Reinvestment Act of 2009 (Recovery Act) ($43.4 million), this was an estimate, and the funding was actually obligated over 2 years ($6 million obligated in fiscal year 2009, and $37.4 million obligated in fiscal year 2010). However, as of February 2012, NIST's estimated data have not been updated on the NNI website.

---

[34]Obligations are definite commitments that create a legal liability of the government for the payment of goods and services ordered or received, or a legal duty on the part of the United States that could mature into a legal liability by virtue of actions on the part of the other party beyond the control of the United States. Payment may be made immediately or in the future. An agency incurs an obligation, for example, when it places an order, signs a contract, awards a grant, purchases a service, or takes other actions that require the government to make payments to the public or from one government account to another. See GAO-05-734SP.

[35]As directed in the President's Memorandum on Transparency and Open Government issued on January 21, 2009, OMB issued an Open Government Directive instructing executive branch agencies to, among other things, publish government information online.

In some instances, the reported data may not reflect all EHS research funding. We found that the fiscal year 2010 data reported by NSTC in the NNI Supplement to the President's 2012 Budget did not include all 2010 nanotechnology funding from Recovery Act appropriations, and OMB did not request that information from agencies in its call for 2010 data.[36] We found that NIH,[37] NIST,[38] and NSF[39] together obligated a total of $131.2 million for nanotechnology research in 2010 with Recovery Act appropriations (including $8 million for EHS research by NIH).[40] However, while NSF included its Recovery Act funds in the funding reported in the NNI Supplement to the President's 2012 Budget, NIH and NIST did not. In addition, some EHS research funding may not be included in the budget supplements because of agency-specific reporting methods.[41]

---

[36]However, the NNI Supplement to the President's 2011 Budget did include data on 2009 actual agency investments from Recovery Act appropriations.

[37]NIH officials told us that, for fiscal year 2010, NIH obligated funds appropriated by the Recovery Act as follows: PCA 1, $19.8 million; PCA 2, $15.6 million; PCA 3, $28.3 million; PCA 4, $3.9 million; PCA 7, $8 million; and PCA 8, $0.5 million.

[38]NIST officials told us that, for fiscal year 2010, NIST obligated funds appropriated by the Recovery Act as follows: PCA 5, $3.1 million; and PCA 6, $34.3 million.

[39]NSF officials told us that, for fiscal year 2010, NSF obligated $17.7 million for PCA 6 research with funds appropriated by the Recovery Act.

[40]We did not determine the amounts of 2010 investments from Recovery Act appropriations, if any, by the NNI member agencies funding nanotechnology research that were not included in this review (the National Aeronautics and Space Administration and the Departments of Defense, Energy, Agriculture, Justice, Homeland Security, and Transportation).

[41]For example, according to NIOSH officials, the agency does not report to OMB funds that provide support for major program areas that require shared services and facilities, because of the difficulty of placing portions of a central support project into a program component area. Also, the automated method NIH uses to compile its list of nanotechnology EHS research projects may not identify all appropriate projects because it relies on a computerized search of key words, which will not define an absolute dividing line between projects that should be included in the EHS program component area and those that should not. According to NIH officials, this process provides a consistent, reliable, and repeatable method for efficiently scanning NIH's projects, which number more than 50,000 annually. We did not review information on NIH projects that were not reported to us by NIH as nanotechnology EHS research. However, in reviewing information on the 58 fiscal year 2010 nanotechnology EHS research projects NIH reported to us, we identified one example of funding for a subproject that was missed by this method.

EHS Research Funded in 2010 Focused On Metal- and Carbon-Based Nanomaterials

In fiscal year 2010, EHS research at the seven NNI member agencies we reviewed focused on two categories of nanomaterials more than others—metal- and carbon-based nanomaterials, which are used in a variety of applications, including electronics, consumer products such as sunscreens, medical products, and protective coatings (see fig. 4). NNI has not prioritized nanomaterials for EHS research, but in October 2011 it outlined criteria for its member agencies to use in doing so. It is not yet clear what effect these criteria will have on how agencies prioritize the nanomaterials they focus on in their EHS research.

Figure 4: Categories of Nanomaterials Studied by Seven NNI Member Agencies' EHS Research Projects, Fiscal Year 2010

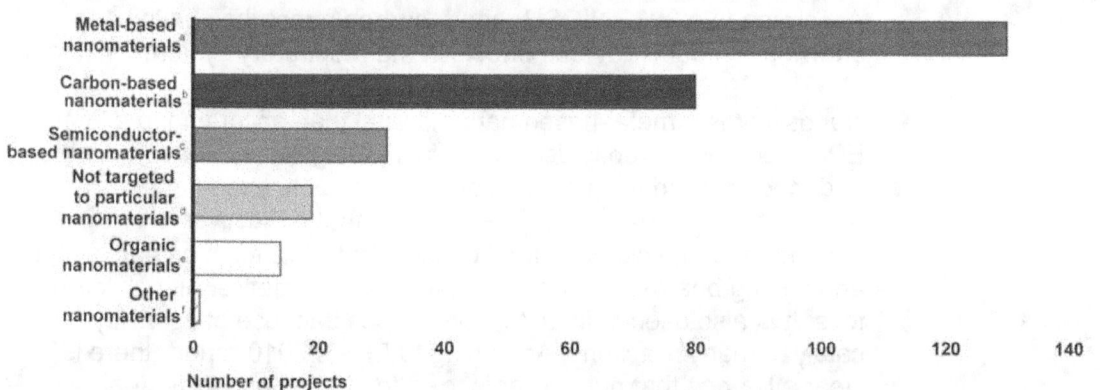

Source: GAO analysis of selected agencies' data on EHS research funded in 2010.

Notes: The figure is based upon our analysis of the 193 projects we determined were primarily directed at EHS. It does not include the 43 projects for which it was not clear to us that the research was primarily directed at EHS. The numbers of projects shown do not total 193 because projects often targeted more than one category of nanomaterials.

[a]Metal-based nanomaterials consisted of silver, gold, iron-based nanomaterials, various metal oxides (titanium, zinc, cerium, and aluminum oxides), and other metal-based nanomaterials.

[b]Carbon-based nanomaterials consisted of carbon nanotubes, fullerenes, and other carbon-based nanomaterials. Spherical and ellipsoidal carbon nanomaterials are referred to as fullerenes, while cylindrical ones are called nanotubes.

[c]Semiconductor-based nanomaterials consisted of quantum dots, silicon-based nanomaterials, and other semiconductor-based nanomaterials. Quantum dots are closely packed semiconductor crystals composed of hundreds or thousands of atoms.

[d]Examples of projects not targeted to particular nanomaterials are projects that funded large research centers on the environmental implications of nanotechnology, or projects that purchased data on the market for nanomaterials in order to guide exposure assessment studies by providing information on potential consumer exposures.

[e]Organic nanomaterials consisted of dendrimers, other polymers, and other organic-based nanomaterials. Dendrimers are branched nanosized polymers. The surface of a dendrimer has numerous chain ends, which can be tailored to perform specific chemical functions.

[f]Other nanomaterials consisted solely of nanoclays (one project).

2010 Research Focused on Carbon Nanotubes, Silver, and Titanium Dioxide

As shown in figure 5, we found that in the 193 fiscal year 2010 research projects we determined were primarily directed at EHS, the nanomaterials that were most frequently the focus of this research were carbon nanotubes (70 projects), nanosilver (60 projects), and nanoscale titanium dioxide (48 projects).[42]

- Carbon nanotubes are nanoscale cylinders consisting of seamlessly "rolled" sheets of graphene, a form of carbon. They are extraordinarily strong, flexible, lightweight, heat resistant, and have high electrical conductivity. Carbon nanotubes are currently used in a variety of applications including conductive coatings for touchscreens and solar cells, and sporting goods such as bicycle frames and baseball bats. According to a 2009 NIOSH report, carbon nanotubes have been shown to produce adverse effects in the respiratory systems of rats.[43]

- Nanosilver is a metal-based nanomaterial that, according to a 2010 EPA report, is currently used in an increasing number of consumer and medical products because of its remarkably strong antimicrobial properties.[44] For example, EPA reported that nanosilver is being incorporated into clothing, food contact materials such as packaging and cutting boards, household appliances, cosmetics, and children's toys. It is also used in industrial processes because of its ability to catalyze many reactions. According to EPA's 2010 report, there is clear evidence that nanosilver is toxic to aquatic and terrestrial organisms and may be detrimental to human health In this report, EPA noted that several studies have shown that nanosilver can be released into the wastewater stream during washing, such as from socks containing nanosilver. The nanosilver released may disrupt the

---

[42]The numbers of projects do not total 193 because (1) some projects targeted nanomaterials other than carbon nanotubes, nanosilver, and nanoscale titanium dioxide; (2) projects often targeted multiple nanomaterials; and (3) nineteen of the 193 projects were not targeted to particular nanomaterials. For example, some projects not targeted to particular nanomaterials funded large research centers on the environmental implications of nanotechnology; others purchased data on the market for nanomaterials to guide the agency's exposure assessment studies by providing information on the potential consumer exposures.

[43]NIOSH, *Approaches to Safe Nanotechnology: Managing the Health and Safety Concerns Associated with Engineered Nanomaterials*, Department of Health and Human Services (NIOSH) Publication No. 2009–125 (March 2009).

[44]EPA, *State of the Science Literature Review: Everything Nanosilver and More*, EPA/600/R-10/084 (August 2010).

helpful bacteria used in wastewater treatment processes or be
released into the environment.

- Nanoscale titanium dioxide is a metal-based nanomaterial used in
sunscreens, protective coatings, and other materials to manage heat
and light by blocking UV light from the sun's rays. It is also being
added to paints, cements, windows, tiles, and other products for its
sterilizing and deodorizing properties and is being used for antifogging
coatings and self-cleaning windows. In addition, it is being
investigated for use in removing contaminants from drinking water.
According to a 2011 NIOSH report, nanoscale titanium dioxide is a
potential occupational carcinogen when inhaled.[45] Regarding
exposure to this nanomaterial in sunscreens, a 2010 FDA publication
found that nanoscale titanium dioxide included in a formulation similar
to currently marketed sunscreens is unlikely to significantly penetrate
human skin.[46]

---

[45]NIOSH, *Current Intelligence Bulletin 63: Occupational Exposure to Titanium Dioxide*, Department of Health and Human Services (NIOSH) Publication No. 2011–160 (April 2011).

[46]N. Sadrieh, A.M. Wokovich, N.V. Gopee, J. Zheng, D. Haines, D. Parmiter, P.H. Siitonen, C.R. Cozart, A.K. Patri, S.E. McNeil, P.C. Howard, W.H. Doub, L.F. Buhse, "Lack of Significant Dermal Penetration of Titanium Dioxide from Sunscreen Formulations Containing Nano- and Sub-Micron-Size TiO$_2$ Particles," *Toxicological Sciences,* vol. 115, no. 1 (2010):156-66.

Figure 5: Nanomaterials Studied by Seven NNI Member Agencies' EHS Research Projects, Fiscal Year 2010

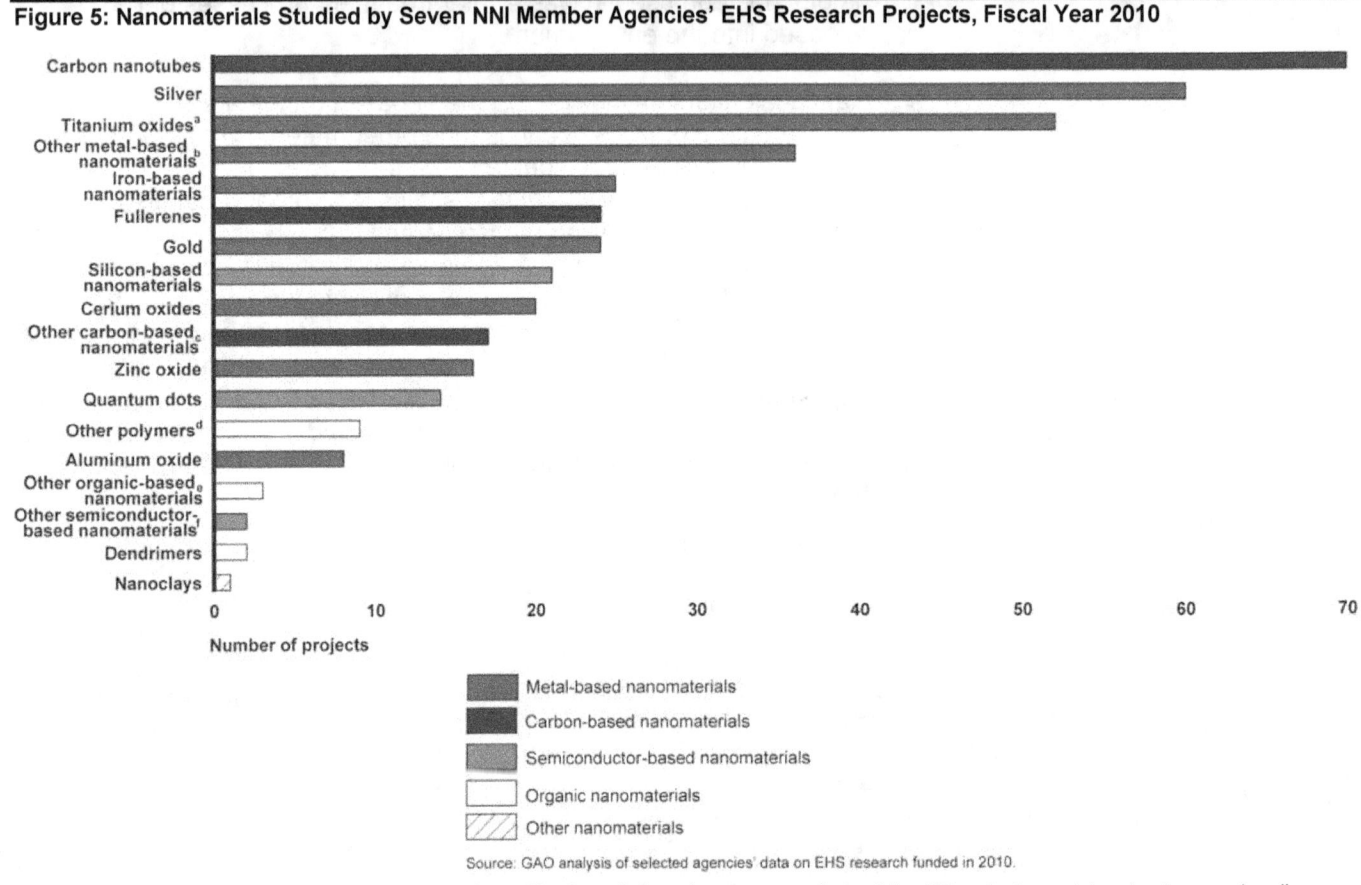

Source: GAO analysis of selected agencies' data on EHS research funded in 2010.

Notes: The figure is based upon our analysis of the 193 projects we determined were primarily directed at EHS. It does not include the 43 projects for which it was not clear to us that the research was primarily directed at EHS, nor those not targeted to particular nanomaterials. The numbers of projects shown do not total 193 because projects often targeted more than one nanomaterial.

[a]Of the 52 projects that targeted titanium oxides, at least 48 projects studied titanium dioxide. Two projects studied titanium oxides, but it was not clear if these included titanium dioxide.

[b]Other metal-based nanomaterials consisted of various metals, such as cobalt, copper, magnesium, manganese, nickel, and others, as well as some composites of multiple metals.

[c]Other carbon-based nanomaterials consisted of carbon nanofibers, carbon nanowires, carbon nanohorns, carbon black, diamond, graphene, nanoscale graphene platelets, plastic composite material containing carbon nanofibers, and soot nanoparticle agglomerates.

[d]Other polymers consisted of polystyrene, polyhedral oligomeric silsesquioxanes, nylon 6 nanofibers, plastics, polymeric nanovesicles, and poly(propargyl glycolid) nanoparticles.

[e]Other organic-based nanomaterials consisted of flourescein, sucrose, and dioctyl phthalate particles.

[f]Other semiconductor-based nanomaterials consisted of selenium nanoparticles and cadmium and lead selenides/sulfides.

Effect of Recent NNI Criteria for Agencies' Prioritization of Nanomaterials Is Not Yet Clear

NNI has not prioritized nanomaterials for EHS research, but the 2011 NNI EHS research strategy outlines criteria to assist its member agencies in doing so. The main criteria NNI proposed are the extent to which (1) study of a particular nanomaterial may provide a major contribution to the research knowledge base, and (2) nanomaterials and nanotechnology-enabled products may pose a safety concern to workers, consumers, and the environment. NNI further identified five criteria for assessing whether a material may pose a safety concern: (1) the nanomaterial's potential for hazard, (2) likelihood of exposure, (3) high reactivity, (4) biological novelty, and (5) the involvement of the material in an event that produced health or environmental impacts. However, because the criteria were published in October 2011, it is too soon to tell how they will influence NNI member agencies' decisions about which nanomaterials to prioritize.

Furthermore, it is unclear if the information needed to use the NNI criteria is available. As we reported in May 2010, predicting and assessing the potential hazards, exposures, and resulting risks from nanomaterials is difficult, and current understanding of nanomaterial toxicity and exposure is limited. For example, the findings from completed toxicity studies of a nanomaterial constructed in one manner may not be applicable to understanding the risks posed by the same nanomaterial constructed in a different manner and, therefore, studies of similar nanomaterials may not be comparable.[47] Also, according to the NRC's 2012 draft strategy for nanotechnology EHS research, there is incomplete information on the effects of the array of nanomaterials used in products and a lack of information on effects of chronic exposures. NRC reported that most toxicity studies test a single material and usually focus on effects of acute exposures.[48]

In addition, reliable, comprehensive information is not readily available on the likelihood of exposure to nanomaterials. Consumers and workers are more likely to be exposed to nanomaterials that are already produced in large quantities or incorporated into a larger number of products, but information on the prevalence or production volumes of nanomaterials or nanotechnology-enabled products is currently limited. As previously noted in this report, FDA, CPSC, and EPA may not receive complete

---

[47]GAO-10-549.

[48]NRC, *A Research Strategy for Environmental, Health, and Safety Aspects of Engineered Nanomaterials* (2012).

information on what nanomaterials are already in use in the case of certain products. CPSC officials told us that one way the agency collects information on the use of nanomaterials in products to better understand potential exposures is by purchasing information from commercial vendors. However, we found that obtaining such information from commercial vendors such as market research companies can be costly, and it is difficult to assess the reliability of such research because it can involve data and methods, including modeling based on various assumptions, that may be proprietary.

Another source of information on nanomaterials in commerce is the Woodrow Wilson International Center for Scholars' Project on Emerging Nanotechnologies, which maintains an inventory of nanotechnology-based products.[49] As of the March 2011 update to the list, the inventory contained approximately 1,300 nanotechnology-based products. However, the Wilson Center's list is not comprehensive, and it consists of consumer products.[50] As we reported in May 2010, occupational exposure is a particular concern because the exposure and risk to workers is potentially greater than the risk to consumers.[51] At present, though, there is little information on the exposure of workers to nanomaterials in the workplace.

---

[49]The Wilson Center is a nonpartisan research institution established by an act of Congress in 1968 and supported by public and private funds. The Project on Emerging Nanotechnologies was established in April 2005 as a partnership between the Wilson Center and the Pew Charitable Trusts. The project's mission is to, among other things, collaborate with researchers, government, industry, and others to identify gaps in knowledge and regulatory processes, develop strategies for closing them, and provide independent, objective knowledge and analysis related to the development and commercialization of nanotechnologies.

[50]According to the Wilson Center, the inventory is based on information that can be readily found on the Internet about products that (1) can be readily purchased by consumers and (2) are identified as nanotechnology-based by the manufacturer or another source. The center tried to avoid including products that clearly do not use nanotechnology, but they did not verify manufacturers' claims or conduct any independent testing of products.

[51]GAO-10-549.

NNI Member Agencies Have Collaborated Extensively on Nanotechnology EHS Research

The NNI member agencies have collaborated extensively with each other and nonfederal stakeholders on EHS research and strategies. The NNI member agencies participated in interagency efforts to develop joint strategies related to EHS research. They have also undertaken over 40 collaborative nanotechnology EHS research projects in recent years, signing agreements with both federal and nonfederal stakeholders. Many of the nonfederal stakeholders responding to our questionnaire rated their collaborative activities with NNI agencies as very or generally useful, although they have identified some challenges.

NNI Member Agencies Have Collaborated to Develop Joint Strategies

NNI member agencies have collaborated through the NSTC to develop joint EHS research strategies. The NSTC primarily coordinates nanotechnology EHS research through the NSET Subcommittee and the NEHI, an interagency working group. The NEHI's purpose includes providing for an exchange of information among agencies and with nonfederal stakeholders; facilitating the identification, prioritization, and implementation of nanotechnology EHS research; and managing the EHS interagency research strategy and facilitating its implementation. The 2011 NNI EHS research strategy was developed by NNI member agencies working through the NEHI working group. The strategy sets a common vision for nanotechnology EHS research, EHS research categories and needs, and key principles intended to assist the NNI member agencies make strategic decisions about research programs that will advance the NNI EHS research agenda while meeting their respective missions. According to several NNI member agency officials, individual agencies' implementation efforts of the 2011 NNI EHS research strategy are discussed at NEHI meetings. However, since the strategy was finalized in October 2011, it is too soon to tell to what extent it will be used to integrate the NNI member agencies' implementation plans. The 2011 NNI EHS research strategy updated and replaced the 2008 NNI *Strategy for Nanotechnology-Related EHS Research*, which had established common priority EHS research needs. The 2008 NNI *Strategy for Nanotechnology-Related EHS Research* was also developed by the NEHI working group and informed by earlier publications of NNI EHS research needs.

Officials from five of the NNI member agencies told us that they use the NEHI working group as a forum to collaborate on nanotechnology EHS research projects. For example, CPSC officials told us that they identify agencies with expertise on testing nanomaterials found in consumer products through discussions in the NSET Subcommittee and NEHI working group, which has led to collaborative testing of exposure and release of selected nanomaterials. This collaboration has resulted in data that support the respective missions of the collaborative agencies. Officials from the National Toxicology Program at NIH noted that they selected nanomaterials for study based in part on discussions in the NEHI working group and NNI sponsored workshops. Officials from NIST told us that NNI member agencies are also working through the NEHI working group to develop an inventory of collaborative EHS research activities. Collaborative activities related to EHS research have been reported in the 2011 NNI EHS research strategy and annual NNI Supplements to the President's Budget.

NNI Member Agencies Have Engaged In Numerous Collaborative EHS Research Projects

The NNI member agencies have recently initiated numerous nanotechnology EHS research projects in collaboration with other federal agencies and nonfederal stakeholders. We reviewed formal collaborative nanotechnology EHS research projects initiated from February 2008 to October 2011, while the 2008 NNI *Strategy for Nanotechnology-Related EHS Research* was in effect. The 2008 NNI *Strategy for Nanotechnology-Related EHS Research* called for the NEHI and the NNI member agencies to coordinate agency efforts to address priority research needs and, among others, identify opportunities for collaboration and joint development and use of resources where appropriate, facilitate partnerships with industry, and coordinate and support international efforts, support development of consensus-based standards, and facilitate wide dissemination of research results and other nonproprietary EHS information. During this period, NNI member agencies initiated 43 formal collaborative projects related to EHS research—24 interagency collaborations and 19 collaborations that included nonfederal stakeholders. Most of the interagency collaborations were among the NNI member agencies included in our review, but a few were with other federal agencies such as the Department of Defense, the Department of Labor's Occupational Safety and Health Administration, and the Department of Agriculture. Nonfederal stakeholders included foreign governments, such as the United Kingdom (UK) and China; universities, such as the University of Massachusetts, Lowell and Rice University; and NGOs, such as the International Alliance for NanoEHS Harmonization

(IANH) and the International Life Sciences Institute Research Foundation.[52]

NNI member agencies used these collaborative projects to extend their capability to achieve their individual missions. For example, CPSC collaborated with four agencies to conduct research on nanomaterials found in consumer products. NIOSH, an agency within the Centers for Disease Control and Prevention that conducts research and makes recommendations for the prevention of work-related injury and illness, partnered with entities that operate nanomanufacturing facilities to better understand the potential for occupational exposure to nanomaterials. NIST, which aims to advance measurement science, standards, and technology, signed an agreement with the IANH to jointly create protocols for toxicological tests of selected representative nanoparticles, confirming inter-laboratory reproducibility and verifying the predictability of certain procedures.

Collaborative EHS research projects have resulted in transfers of funding as well as sharing of expertise, facilities, and other resources, as shown in the following examples:

- An interagency agreement between CPSC and NIOSH to study the pulmonary effects of titanium dioxide nanoparticles released from aerosol spray products involved CPSC providing the product to be tested and transferring funding to NIOSH to construct the testing equipment, with NIOSH providing expertise and staff time to run the tests and produce a report.

- An agreement among the NIH's National Cancer Institute, the National Institute of Environmental Health Sciences (NIEHS), and the National Institute of Biomedical Imaging and Bioengineering to develop, maintain, and operate a web-based nanomaterial registry specified that the National Institute of Biomedical Imaging and Bioengineering will develop the registry to provide consistent and curated information on the biological and environmental interactions of

---

[52]The IANH is a group of internationally recognized experts who have agreed to develop specific tools and testing protocols and perform a set of round robin experiments for reproducible testing of nanomaterial biological interactions and toxicology. The International Life Sciences Institute Research Foundation is a nonprofit organization based in Washington, D.C., that focuses on advancing the methods and application of science in risk assessment, human nutrition, and the prevention of obesity.

well-characterized nanomaterials, as well as links to associated publications, modeling tools, computational results, and manufacturing guidance from existing databases. The National Cancer Institute and the NIEHS provided funding. NIEHS also committed to provide input to the constitution of an advisory board and attend its meetings, as well as be represented in the nano registry project management team.

- Agencies have also issued joint solicitations for grant applications, such as one between EPA and a number of UK agencies to solicit proposals for research on environmental and health implications of nanotechnology. EPA and the UK agencies signed an agreement that notes (1) the funding of grants will be made consistent with the budget priorities of each party and (2) they are to work jointly to define the scientific priorities, issue the request, assist with peer review and selection of grantees, and disseminate the research results.

Details on these and other collaborative EHS research projects are provided in appendix II.

Nonfederal Stakeholders We Surveyed Benefited from NNI Collaboration but Identified Some Challenges

The 138 nonfederal stakeholders who responded to our questionnaire reported that they benefited from collaboration with NNI member agencies but faced some challenges. Most respondents indicated that the 2011 NNI EHS research strategy was very or somewhat effective at addressing the EHS research needs. Respondents reported that they also obtained information on nanotechnology EHS risks from NNI member agencies and nongovernmental sources.

Many of the Respondents to Our Questionnaire Rated Collaboration with NNI Member Agencies on EHS Research as Generally or Very Useful

The following three formal mechanisms for collaboration with NNI member agencies were most frequently identified by respondents as generally or very useful to them: (1) joint data gathering and sharing; (2) joint research solicitations or funding of research consortia; and (3) competitive grants. See figure 6 for respondents' ratings of seven formal collaborative mechanisms we identified for them to rate. Some respondents who provided comments in response to optional open-ended questions also identified public-private partnerships, such as joint academic, government, and industry information exchange and research programs, as collaboration mechanisms the NNI member agencies should consider. A few respondents who commented in response to optional open-ended questions also cited benefits to informal collaboration with NNI agencies, such as discussions during workshops and conferences.

Figure 6: Usefulness of Formal Collaboration Mechanisms for Nanotechnology EHS Research, According to Questionnaire Respondents

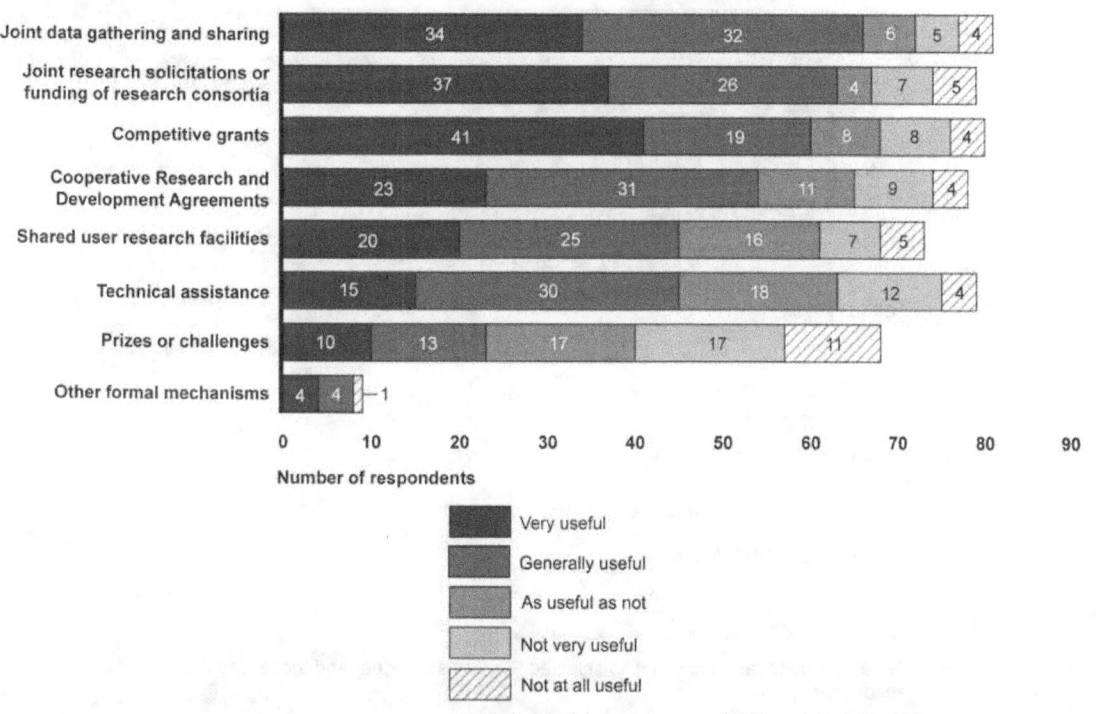

Number of respondents

■	Very useful	
■	Generally useful	
■	As useful as not	
☐	Not very useful	
▨	Not at all useful	

Source: GAO analysis of questionnaire of nonfederal stakeholders.
Note: Excludes respondents who selected "No basis to judge" and those who did not check a response.

Since both the 2011 NNI EHS research strategy and its predecessor, the 2008 NNI *Strategy for Nanotechnology-Related EHS Research*, called for partnerships with industry and other nonfederal stakeholders, we included questions about collaboration with NNI member agencies in our questionnaire. When asked about the usefulness of their collaboration with the NNI member agencies, more than half of respondents rated their collaborative EHS research or related activities for each of the NNI member agencies as generally or very useful, as seen in figure 7.

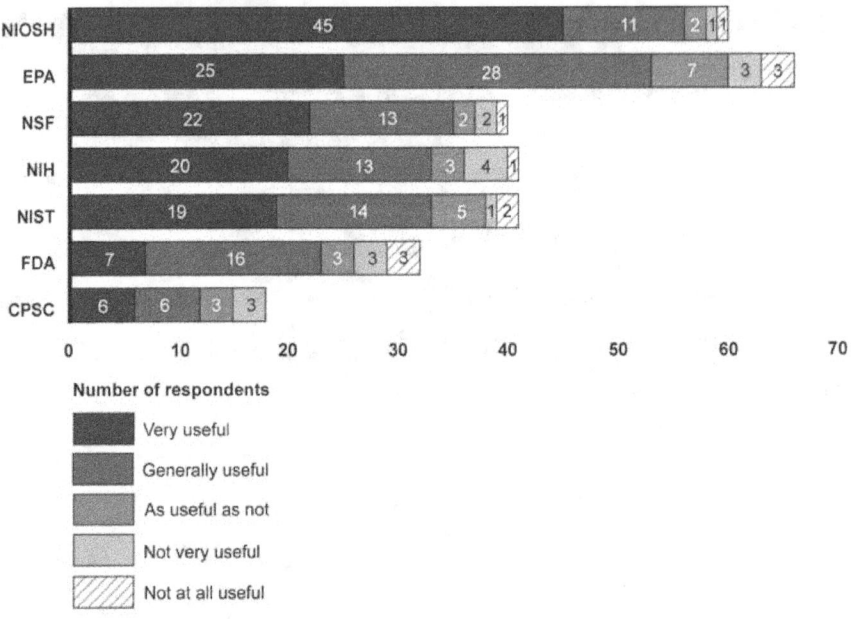

Figure 7: Usefulness of Collaborative EHS Research or Related Activities with NNI Member Agencies, According to Questionnaire Respondents

Source: GAO analysis of questionnaire of nonfederal stakeholders.

Note: Excludes respondents who selected "No basis to judge" and those who did not check a response.

Respondents to Our Questionnaire Indicated a Number of Challenges to Collaboration with NNI Member Agencies Apply to Them

Stakeholders responding to our questionnaire indicated that certain challenges to collaboration with NNI member agencies we identified on EHS research apply to them. As seen in figure 8, more than two-thirds of those who rated the challenge of lack of funding and the challenge of limited awareness of collaboration opportunities indicated that each of these challenges apply to them. This is consistent with what we heard from representatives of a nanotechnology trade association and an NGO, who stated that there is limited funding for corporate EHS research and development efforts because they do not generate revenues. Some respondents who provided comments in response to optional open-ended questions stated that the private sector needs government funding for EHS research. A number of respondents also commented that government funding for EHS research was inadequate or an "after-thought" to other purposes. Some respondents also commented that funding seems to be targeted toward large, multiyear centers rather than smaller targeted projects that can help address near-term EHS needs.

More than half of those responding noted that each of the challenges of regulatory uncertainty, lack of standardization—such as different terminology or protocols—and concerns regarding disclosure of proprietary or confidential business information, apply to them. Some respondents cited regulatory uncertainty—that is, lack of a clear regulatory environment to enable commercialization and protect consumers and the general public—as a challenge to collaboration with their work with NNI member agencies on nanotechnology EHS research. For example, a number of respondents who provided comments in response to optional open-ended questions pointed to regulatory uncertainty as limiting the ability of companies to determine their EHS responsibilities. Some respondents also commented on difficulties in collaborating with NNI member agencies resulting from differences in expertise and regulatory approaches across different government agencies or even within the same agency. With regard to lack of standardization as a challenge to collaboration on nanotechnology EHS research, some respondents provided examples related to the lack of scientific consensus on the definition of nanomaterials, the testing methods for toxicology research, and common terminology across research and regulatory agencies.

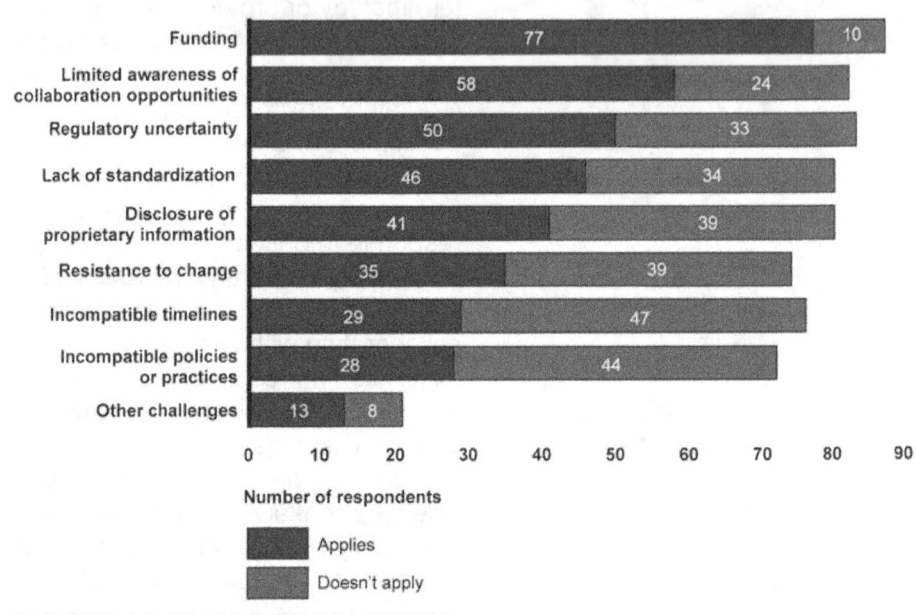

Figure 8: Challenges to Collaboration on Nanotechnology EHS Research, According to Questionnaire Respondents

Source: GAO analysis of questionnaire of nonfederal stakeholders.

Note: Excludes respondents who selected "Don't know/Not sure" and those who did not check a response.

Most Respondents to Our Questionnaire Rated the 2011 NNI EHS Research Strategy as Somewhat or Very Effective

When asked to rate the effectiveness of mechanisms we identified that NNI member agencies can use for obtaining input from nonfederal stakeholders on the development of the NNI EHS research strategies, respondents most frequently identified the following three mechanisms as being used somewhat effectively or effectively: (1) workshops, conferences, and other public speaking engagements; (2) advisory councils; and (3) the NNI web portal.

Figure 9: Effectiveness of Mechanisms for Obtaining Input on the NNI Strategic Planning for EHS Research, According to Questionnaire Respondents

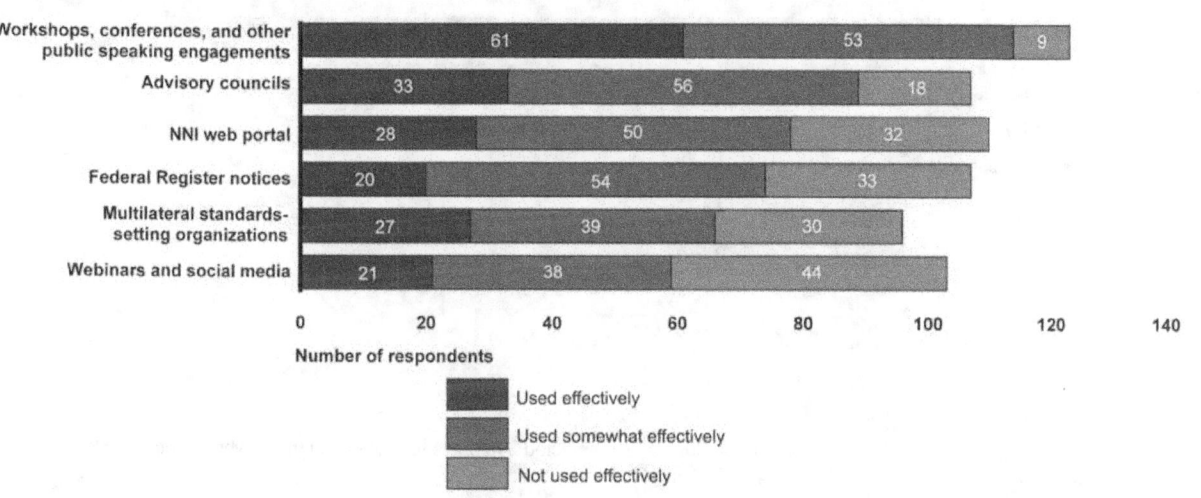

Number of respondents

Used effectively

Used somewhat effectively

Not used effectively

Source: GAO analysis of questionnaire of nonfederal stakeholders.

Note: Excludes respondents who selected "No basis to judge" and those who did not check a response.

When asked to rate the 2011 NNI EHS research strategy, most of those responding rated this strategy as somewhat or very effective at addressing nanotechnology EHS research needs. As seen in figure 10, while just over half of those responding to these questions indicated the 2008 NNI *Strategy for Nanotechnology-Related EHS Research* addressed these needs somewhat or very effectively, 79 out of 98 indicated the 2011 NNI EHS research strategy addressed these needs somewhat or very effectively.[53] Some of the respondents who commented in response to optional open-ended questions commended the 2011 NNI EHS research strategy for accurately capturing the input of the workshops' participants and summarizing the research needs, but others pointed to shortcomings, including a lack of prioritization and insufficient focus on implementation. For example, one respondent commented that the strategy would only be effective if it was implemented, and that it may not be sufficient to rely on the agencies' voluntary implementation efforts.

[53]The NSTC finalized the 2011 NNI EHS research strategy while the questionnaire was being completed by respondents. Therefore, some respondents may have rated the draft 2011 NNI EHS research strategy while others may have rated the final version.

Figure 10: Effectiveness of NNI EHS Research Strategies at Addressing Nanotechnology EHS Research Needs, According to Questionnaire Respondents

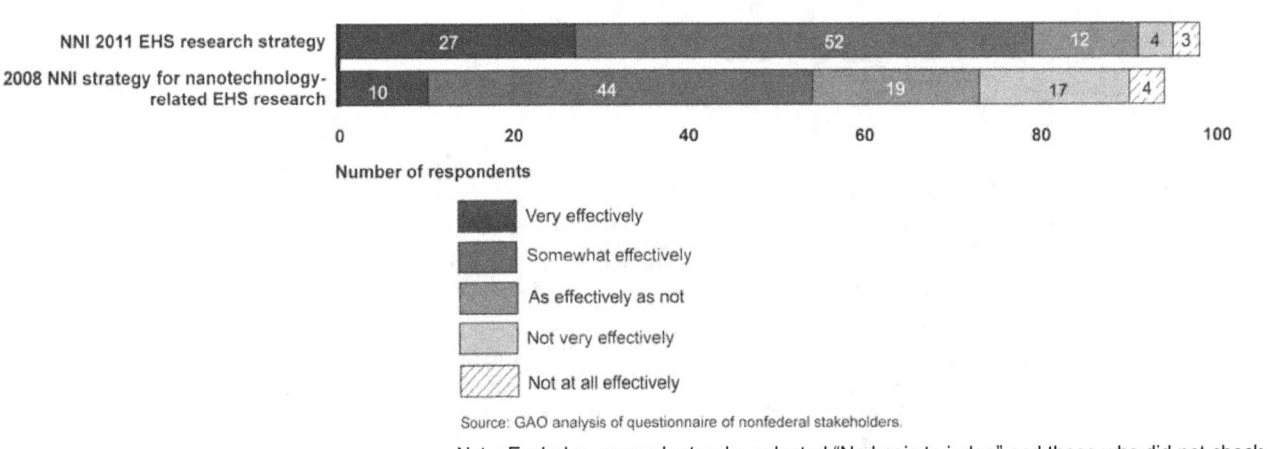

Number of respondents

Very effectively

Somewhat effectively

As effectively as not

Not very effectively

Not at all effectively

Source: GAO analysis of questionnaire of nonfederal stakeholders.

Note: Excludes respondents who selected "No basis to judge" and those who did not check a response.

Respondents Obtained Nanotechnology EHS Information from Various Sources

Additionally, respondents identified various sources, including the NNI member agencies, from which they obtained information on the potential EHS risks of nanotechnology. When asked to estimate the frequency with which they obtained nanotechnology EHS information from NNI member agencies in recent years, respondents reported obtaining information from the NNI member agencies in varying frequencies. More than half of the respondents to these questions reported occasionally or often obtaining information from EPA and NIOSH, as seen in figure 11.

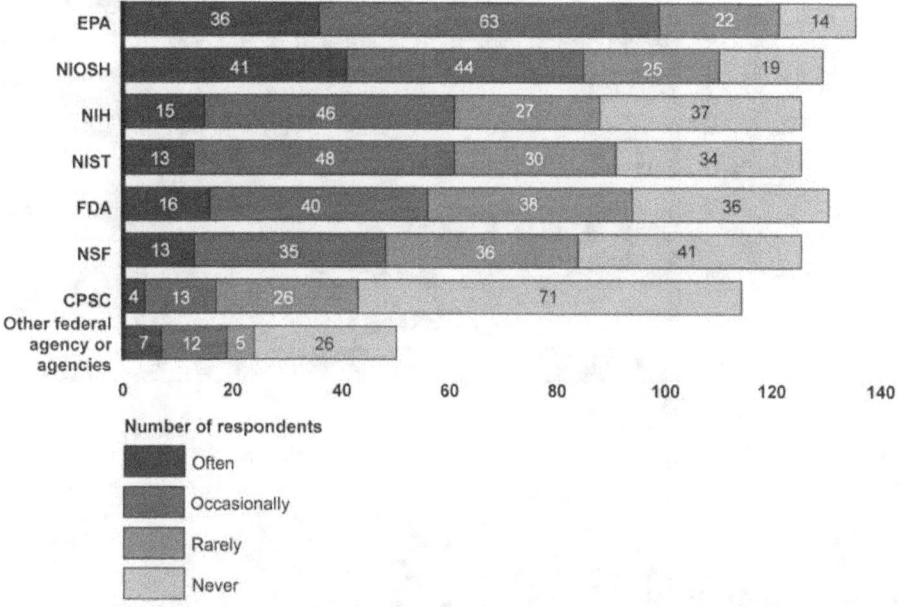

Figure 11: Frequency with Which Questionnaire Respondents Reported Obtaining Information on the Potential EHS Risks of Nanotechnology from NNI Member Agencies

Source: GAO analysis of questionnaire of nonfederal stakeholders.

Note: Excludes respondents who selected "Don't know/Not sure" and those who did not check a response.

The 2011 NNI strategic plan calls for providing a "one-stop" Internet-based portal for nanotechnology information, including, among other things, scientific data such as characterization and toxicity measurements. About 80 percent of the 131 stakeholders responding to a question about how useful such a portal would be indicated that an Internet-based portal for nanotechnology EHS information would be generally or very useful to them. The NNI does not currently provide a portal that contains this information, but an NNCO official told us that the NNCO is implementing a web-based map that identifies the instrumentation and facilities for nanotechnology research at major universities and centers in the U.S. and federal labs.

When asked to rate the frequency with which they obtained information on the potential EHS risks of nanotechnology from nongovernmental sources, as seen in figure 12, more than two-thirds of respondents reported that they occasionally or often obtained information from each of the following four sources: (1) peer-reviewed scientific publications; (2) in-house research; (3) online databases; and (4) news outlets.

Figure 12: Frequency with Which Questionnaire Respondents Reported Obtaining Information on the Potential EHS Risks of Nanotechnology from Nongovernmental Sources

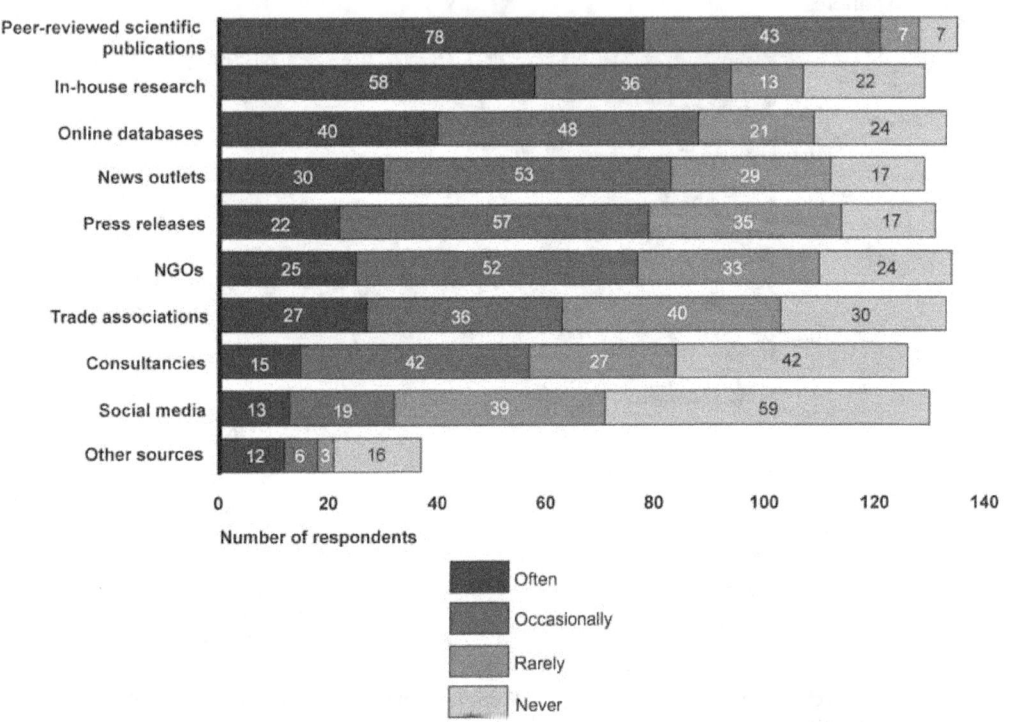

Source: GAO analysis of questionnaire of nonfederal stakeholders.

Note: Excludes respondents who selected "Don't know/Not sure" and those who did not check a response.

NNI Strategy Documents Address or Partially Address Desirable Characteristics of National Strategies

NNI strategy documents address two and partially address four of the six desirable characteristics of effective national strategies that we identified in prior work and that offer policymakers and implementing agencies a management tool that can help ensure accountability and more effective results. The three NNI strategy documents—the 2011 NNI strategic plan, the 2011 NNI EHS research strategy, and the NNI Supplement to the President's 2012 Budget[54]—compose a national strategy for nanotechnology EHS research and create a framework for achieving NNI program goals and priorities.

We rated the NNI strategy documents to determine how well they jointly addressed the six characteristics that we have identified for effective national strategies, as described in table 3.

Table 3: Summary of Desirable Characteristics for a National Strategy

Desirable characteristic	Brief description
Purpose, scope, and methodology	Addresses why the strategy was produced, the scope of its coverage, and the process by which it was developed.
Problem definition and risk assessment	Addresses the particular national problems and threats the strategy is directed toward.
Goals, subordinate objectives, activities, and performance measures	Addresses what the strategy is trying to achieve; steps to achieve those results; as well as the priorities, milestones, and performance measures to gauge results.
Resources, investments, and risk management	Addresses what the strategy will cost, the sources and types of resources and investments needed, and where resources and investments should be targeted by balancing risk reductions and costs.
Organizational roles, responsibilities, and coordination	Addresses who will be implementing the strategy, what their roles will be compared to others, and mechanisms for them to coordinate their efforts.
Integration and implementation	Addresses how a national strategy relates to other strategies' goals, objectives, and activities—and to subordinate levels of government and their plans to implement the strategy.

Source: GAO-04-408T.

---

[54]NSTC, Committee on Technology, Subcommittee on Nanoscale Science, Engineering, and Technology, *National Nanotechnology Initiative Environmental, Health, and Safety Research Strategy* (October 2011); *National Nanotechnology Initiative Strategic Plan* (February 2011); and *The National Nanotechnology Initiative Supplement to the President's 2012 Budget* (February 2011).

Ideally, effective national strategies should fully address all of these characteristics. However, we recognize that by their nature, national strategies are intended to provide broad direction and guidance—rather than be prescriptive, detailed mandates—to the relevant implementing agencies. Thus it is unrealistic to expect all national strategies to provide details on each and every characteristic we identified. Moreover, the NNI member agencies have different statutory authority and functions that may significantly affect their research and research priorities. Nonetheless, we believe the more detail a strategy provides, the easier it is for the responsible parties to implement it and achieve its goals.

We reviewed the three NNI strategy documents for elements related to these characteristics, and based on the inclusion of these elements, rated how well the strategic documents address the six characteristics. According to our methodology, the strategy documents "address" a characteristic when they explicitly include all, or nearly all, elements of the characteristic and have sufficient specificity and detail. The documents "partially address" a characteristic when they include some or most of the elements of the characteristic with sufficient specificity and detail.[55] The documents "do not address" a characteristic when they do not include any elements of a characteristic or references are too vague or general to be useful. Additional details about our ratings and the elements that make up these characteristics are available in appendix I.

As described in table 4, we found, when reviewed as a whole, that the strategy documents address or partially address all of the desirable characteristics of a national strategy.

---

[55]The "partially addresses" category includes a range that varies from explicitly citing most of the elements to citing as few as one of the elements of a characteristic. Yet because the three NNI documents have different purposes and audiences an individual document need not address each desirable national strategy characteristic. However, areas not addressed in one document should be more fully developed in another.

Table 4: Extent NNI Strategy Documents Address GAO's Desirable Characteristics with Respect to Nanotechnology EHS Research

Desirable characteristic	Address	Partially address	Do not address
1) Purpose, scope, and methodology	X		
2) Problem definition and risk assessment	X		
3) Goals, subordinate objectives, activities, and performance measures		X	
4) Resources, investments, and risk management		X	
5) Organizational roles, responsibilities, and coordination		X	
6) Integration and implementation		X	

Source: GAO analysis of the NNI strategy documents.

Purpose, Scope, and Methodology and Problem Definition and Risk Assessment (Desirable Characteristics 1 and 2)

Taken as a whole, the NNI strategy documents address the first two characteristics—purpose, scope, and methodology and problem definition and risk assessment—by including all, or nearly all, of their elements. These characteristics describe the scope of the strategy, describe how and why it was produced, define the problems the strategy intends to address, discuss causes and the operating environment, and provide a risk assessment or broad description of potential risks. Elements of these characteristics that are included in the documents are providing a clear statement of purpose, defining key terms, delineating what major functions or mission areas are covered, discussing problems with the current understanding of EHS implications of nanotechnology, and the quality of currently available data, among others.

Goals, Subordinate Objectives, Activities, and Performance Measures (Desirable Characteristic 3)

The NNI strategy documents partially address the third characteristic regarding goals, subordinate objectives, activities, and performance measures. This characteristic describes the overall desired results, hierarchy of strategic goals and subordinate objectives, priorities, milestones, outcome-related performance measures, and process for monitoring and reporting on progress, among others. The NNI strategy documents include a number of elements such as desired results, strategic goals, and subordinate objectives, but do not include, or do not fully develop, other elements such as priorities, milestones, or outcome-related performance measures. Performance information—such as outcome-related performance measures and milestones comprised of targets and time frames for meeting those measures—allows managers to identify performance problems and develop approaches that improve results. For the purposes of this report, we define performance information to mean data collected to measure progress toward achieving an established goal. A wide range of information can be relevant to program performance. Performance information can focus on various dimensions of performance such as outcomes, outputs, quality, timeliness, customer satisfaction, or efficiency. It can inform key management decisions such as setting program priorities, allocating resources, identifying program problems, and taking corrective action to solve those problems; or it can help determine progress in meeting the goals of programs or operations. The NNI strategy documents contain detailed research needs for nanotechnology EHS and report the annual funding of EHS research and the number of projects supported for selected years, but the documents do not prioritize among these needs or include outcome-related performance measures, targets, or time frames that allow for monitoring and reporting on progress toward meeting the research needs.

Independent reviews of the prior NNI strategy documents also noted an absence of performance information. In 2010, the National Nanotechnology Advisory Panel recommended that the NNCO monitor metrics that assess, among other things, the NNI's progress on developing methodologies to assess plausible risks of nanotechnology. The advisory panel also recommended that the NEHI working group develop and implement a strategy that links EHS research activities with knowledge gaps and decision-making needs. The 2009 NRC review concluded that the 2008 NNI *Strategy for Nanotechnology-Related EHS Research* could be an effective tool, but did not include a number of elements, including measures of research progress. We previously reported that, ideally, a national strategy would set clear desired results and priorities, outcome-related performance measures, and specific milestones while giving implementing parties flexibility to pursue and achieve those results within a reasonable time frame.[56]

The NNI strategy documents give agencies wide latitude to develop research programs to meet the goals and research needs of the NNI, but we found that NNI member agencies vary in their identification and reporting of agency-specific performance information for nanotechnology

---

[56]GAO-04-408T.

EHS research that could align with the NNI research needs. The 2011 NNI EHS research strategy states that prioritization, timing, and staging of the research needs identified by the strategy are components of an implementation plan and should be developed within agency missions and appropriations. Similarly, the 2011 NNI strategic plan states that the document serves as a guide for individual agency implementation. Six of the seven NNI agencies we reviewed have documented performance measures, targets, or time frames for their EHS research. Three of the NNI member agencies—CPSC, FDA, and NIH—annually report EHS-related performance information through their publicly available annual performance reports.[57] Three other NNI member agencies—EPA, NIOSH, and NIST—identify strategic goals and performance measures and targets in their program-level nanotechnology EHS strategies, but have not collected, or have not publically reported results of these performance measures.[58] NIOSH has previously published research summaries in its nanotechnology research progress reports. NIOSH officials told us that a forthcoming progress report will also document the results of performance measures identified in their nanotechnology EHS strategy. NSF reports planned EHS research activities as part of its annual budget request, but does not identify performance targets or measures for its nanotechnology EHS research beyond funding requested. According to an NNCO official, the NEHI working group is piloting an effort to gather information from the NNI member agencies to assess high-level progress toward meeting NNI research needs, but has not released the results of this work.

Resources, Investments, and Risk Management and Organizational Roles, Responsibilities, and Coordination (Desirable Characteristics 4 and 5)

The NNI strategy documents partially address the fourth and fifth characteristics describing resources, investments, and risk management and organizational roles, responsibilities, and coordination. We found that the documents include some, but not all, of the elements of these characteristics. With respect to the fourth characteristic, we found that the documents include elements relating to agency resources associated with the strategy such as descriptions of current activities of NNI member

---

[57]CPSC, *2010 Performance and Accountability Report* (November 2010); Department of Health and Human Services, Food and Drug Administration, *FY2012 Online Performance Appendix*; Department of Health and Human Services, National Institutes of Health, *FY2012 Online Performance Appendix*.

[58]EPA, *Nanomaterial Research Strategy*, EPA 620/K-09/011 (June 2009); NIOSH, *Strategic Plan for NIOSH Nanotechnology Research and Guidance*, Department of Health and Human Services (NIOSH) Publication No. 2010-105 (November 2009); NIST, *Nanomaterial Environmental, Health and Safety (Nano-EHS) Program Plan, FY10-FY14*.

agencies and current investment levels. However, we found that the documents do not include, or do not fully develop, other elements such as the costs and types of resources—for example human capital or research and development costs—associated with implementing the strategy. An NNCO official told us that the NNI does not analyze the costs of resources to meet the EHS research needs because it does not control the funding for member agencies, and that each agency allocates funding and prioritizes its activities based on its mission area and available budget. According to this official, the 2011 NNI EHS research strategy identifies research goals, but it is up to the agencies to determine how their funding should be spent. Similarly, a NIST official told us that the NNI cannot dictate to agencies how or what research to fund. With respect to the fifth characteristic pertaining to organizational roles, responsibilities, and coordination, we found that the documents include elements such as lead, support, and partner roles for agencies and processes for coordination and collaboration, but do not include, or do not fully develop, other elements such as an accountability and oversight framework. The strategy documents include detailed descriptions of the NNI structure, describe coordination mechanisms such as the NEHI working group, and provide examples of collaborative activities related to EHS research. As we reported above, we found that NNI member agencies have engaged in numerous collaborative projects with federal and nonfederal partners and nonfederal partners generally rated these collaborations as generally or very useful. However, the strategy documents do not provide details on oversight of agency roles or how agencies will be held accountable to the goals and research needs of the NNI strategy documents.

Integration and Implementation (Desirable Characteristic 6)

The NNI strategy documents also partially address the sixth characteristic describing integration and implementation. This characteristic describes how the national strategy documents relate to subordinate levels of government and their plans to implement the strategy. As we have previously reported, to achieve a common outcome, collaborating agencies need to establish strategies that work in concert with those of their partners or are joint in nature.[59] We found that the NNI strategy documents include limited information regarding the elements of this characteristic. For example, the 2011 NNI EHS research strategy

---

[59]GAO, *Results-Oriented Government: Practices That Can Help Enhance and Sustain Collaboration Among Federal Agencies*, GAO-06-15 (Washington, D.C.: Oct. 21, 2005).

describes a framework for implementation, but does not provide detailed implementation guidance.[60] In addition, we found that the NNI strategy documents generally do not include information on the NNI member agencies' nanotechnology EHS strategies or plans or their integration with the national strategy.

Five of the seven NNI member agencies we reviewed have developed such agency-level nanotechnology EHS research strategies. As noted above, EPA, NIOSH, and NIST have developed strategic plans specifically for EHS research. In addition, FDA has published a strategic plan for its regulatory science research efforts, which include nanotechnology, and developed a research plan specifically targeting nanotechnology.[61] Furthermore, NSF officials told us that the agency describes its strategic direction for nanotechnology, including current and proposed activities and funding by PCA, in its annual budget request.[62] CPSC includes nanotechnology as an external risk factor for its agencywide strategic plan, but the plan does not address EHS research directly. NIH officials told us they do not have agency or institute-specific strategic plans that specifically address nanotechnology EHS research, but nanotechnology EHS research plans developed by its institutes conform to the NNI strategy. Of the five agencies with nanotechnology EHS-related strategies, two—FDA and NIOSH—explicitly align agency research goals in their strategic plans for nanotechnology to NNI EHS research needs. The FDA and NIOSH strategic plans describe the agencies' EHS research goals, their framework for meeting those goals, and which NNI EHS research needs will be advanced. NIST's Nano-EHS program plan, EPA's Nanomaterial Research Strategy, and NSF's budget request to Congress describe agency research objectives. The documents do not explicitly identify NNI research needs; however, agency officials told us that they reflect the NNI strategic documents. The 2011 NNI strategic plan reports that NNI member agencies have initiated

---

[60]The 2011 NNI EHS research strategy states that, because individual agency priorities may differ in scope and focus from the NNI research needs, the NSTC relies on coordination through NEHI to ensure integration of agency plans.

[61]FDA, *Advancing Regulatory Science at FDA* (August 2011), http://www.fda.gov/ScienceResearch/SpecialTopics/RegulatoryScience/ucm267719.htm, and FDA, *Nanotechnology Regulatory Science Research Plan*, http://www.fda.gov/ScienceResearch/SpecialTopics/Nanotechnology/ucm273325.htm.

[62]NSF, *NSF FY2012 Budget Request to Congress*, http://www.nsf.gov/about/budget/fy2012/index.jsp.

a mapping exercise to evaluate how this strategic plan relates to NNI member agencies' strategic plans and to the priorities of the administration. In addition, according to NNI member agency officials, individual agencies' implementation efforts of the 2011 NNI EHS research strategy are discussed at NEHI meetings. However, the results of this mapping exercise and proceedings of the NEHI are not publically available.

Conclusion

The increasing commercialization of nanotechnology-enabled products and information gaps related to nanotechnology EHS impacts underscore the importance of ensuring progress toward meeting EHS strategic goals and research needs. The NNI strategy documents help shape the policies, programs, priorities, resource allocations, and standards related to nanotechnology EHS research and activities. The NNI strategy documents address or partially address the desirable characteristics of effective national strategies, including overarching strategic goals and objectives such as the EHS research needs. However, the documents do not include some elements of the characteristics, which could make it more difficult for federal agencies and other stakeholders to implement the strategy and achieve the identified results. For example, progress toward achieving the strategy documents' goals and research needs cannot be tracked because the NNI does not report specific performance information, such as performance measures, targets, or time frames. In addition, the NNI strategy documents do not include estimates of the costs and types of resources associated with these goals and research needs.

Developing and coordinating performance information and cost estimates for the NNI's nanotechnology EHS research may be challenging because of the varied missions and priorities of the NNI member agencies. However, not having performance information that is aligned with the strategic goals and research needs of the NNI makes it difficult for agencies, policymakers, and stakeholders to determine the collective progress of the national nanotechnology program. Furthermore, developing performance information and including it in NNI strategy documents could help to strengthen two other desirable characteristics that the documents partially address. Specifically, this information could also support an accountability and oversight framework—that is currently not well-developed—by linking agency activities to specific measures of performance and creating consistent benchmarks to judge the overall progress of the NNI member agencies. In addition, developing this information could facilitate better integration between the NNI strategy documents and the NNI member agencies' nanotechnology EHS research strategies and goals. Providing cost estimates related to the NNI EHS research needs would allow the NNI member agencies, policymakers, and stakeholders to assess if investments are commensurate with costs of the identified needs. The NSTC, which is administered by the OSTP, does not direct funding to meet the NNI research needs; however, rough estimates of their costs could serve as guidance to the NNI member agencies as they individually determine their own budgets and priorities. In addition, the NSTC is well-positioned to develop these estimates due to its coordinating role across the member agencies of the NNI and corresponding access to expertise at these agencies. Developing performance information and cost estimates could also support the analysis of the progress made toward achieving the goals and priorities established for the program, as required by the 21st Century Nanotechnology Research and Development Act.

In addition, we are reiterating our 2008 recommendation that the Director of OSTP, in consultation with the Directors of the NNCO and OMB, provide better guidance to agencies on how to report their nanotechnology EHS research.[63] The continued absence of detailed guidance on how to report research under PCA 7 has contributed to the data reporting problems we identified. Therefore, agencies, policymakers, and stakeholders do not have access to accurate, consistent, and complete data on the federal government's investment in nanotechnology EHS research. Without reliable and up-to-date data, it will be difficult for agencies to accurately assess and report their progress toward their own performance measures, as well as the EHS research goals and needs identified in the 2011 NNI EHS research strategy.

Recommendations for Executive Action

To better offer policymakers and implementing agencies a management tool that can help ensure accountability and more effective results, we are making two recommendations to the Director of OSTP:

We recommend that the Director of OSTP coordinate development by the NNI member agencies of performance measures, targets, and time frames for nanotechnology EHS research that align with the research needs of the NNI, consistent with the agencies' respective statutory authorities, and include this information in publicly available reports.

---

[63]GAO-08-402.

We also recommend that, to the extent possible, the Director of OSTP coordinate the development by the NNI member agencies of estimates of the costs and types of resources necessary to meet the EHS research needs.

Agency Comments and Our Evaluation

We provided a draft of this report to the Director, OSTP; Secretary, Commerce; Chairman, CPSC; Administrator, EPA; Secretary, Health and Human Services; and Director, NSF. OSTP and the agencies neither agreed nor disagreed with the recommendations. OSTP and Health and Human Services provided technical and clarifying comments, which we incorporated as appropriate. The Department of Commerce, CPSC, EPA, and NSF indicated they had no comments on the report.

As agreed with your office, unless you publicly announce the contents of this report earlier, we plan no further distribution until 30 days from the report date. At that time, we will send copies to the Director, OSTP; Secretary, Commerce; Chairman, CPSC; Administrator, EPA; Secretary, Health and Human Services; Director, NSF; and other interested parties. The report also will be available at no charge on the GAO website at http://www.gao.gov.

If you or your staff have any questions about this report, please contact me at (202) 512-3841 or ruscof@gao.gov. Contact points for our Offices of Congressional Relations and Public Affairs may be found on the last page of this report. GAO staff who made major contributions to this report are listed in appendix III.

Sincerely yours,

Frank Rusco
Director, Natural Resources
  and Environment

Appendix I: Objectives, Scope, and Methodology

Our review examines (1) changes in federal funding for nanotechnology environmental, health, and safety (EHS) research from fiscal years 2006 to 2010; (2) the nanomaterials that National Nanotechnology Initiative (NNI) member agencies focused on in their EHS research in fiscal year 2010; (3) the extent to which NNI member agencies collaborate with stakeholders on nanotechnology EHS research and related strategies; and (4) the extent to which NNI strategy documents address desirable characteristics of national strategies.

To conduct this work, we reviewed EHS research efforts funded by seven NNI member agencies, which collectively funded 93 percent of EHS research dollars in fiscal year 2010: the National Science Foundation (NSF), National Institutes of Health (NIH), Environmental Protection Agency (EPA), National Institute for Occupational Safety and Health (NIOSH), Food and Drug Administration (FDA), National Institute of Standards and Technology (NIST), and the Consumer Product Safety Commission (CPSC). The first six of these agencies represent the top six providers of EHS research funding from fiscal year 2006 to fiscal year 2010, and CPSC has an important role in ensuring the safe use of nanotechnology in consumer products.

To examine recent changes in federal funding for nanotechnology EHS research, we reviewed information published by the National Science and Technology Council (NSTC) in the NNI Supplements to the President's Budget. Specifically, we reviewed the actual agency investments reported in each program component area for fiscal years 2006 through 2010 for all NNI member agencies funding nanotechnology research.[1] For the dollar amounts that we adjusted for inflation, we used the Biomedical Research and Development Price Index to report funding in constant 2010 dollars. We consulted with the Office of Management and Budget (OMB) and officials at the seven selected agencies to determine the type of budget information reported as actual agency investments in these documents. For fiscal year 2010, the most recent year for which actual

---

[1]See NSTC, Committee on Technology, Subcommittee on Nanoscale Science, Engineering, and Technology, *The National Nanotechnology Initiative Supplement to the President's 2012 Budget* (February 2011); *The National Nanotechnology Initiative Supplement to the President's 2011 Budget* (February 2010); *The National Nanotechnology Initiative Supplement to the President's 2010 Budget* (May 2009); *The National Nanotechnology Initiative Supplement to the President's 2009 Budget* (September 2008); and *The National Nanotechnology Initiative Supplement to the President's 2008 Budget* (July 2007).

agency investment data were available, we also collected quantitative
and qualitative project-level data (such as project funding amounts and
project abstracts or progress reports) on all research projects that the
seven selected agencies had categorized as Program Component Area
(PCA) 7 (EHS research). The agencies do not report project-level data to
OMB annually but report their total EHS research funding to OMB
annually for inclusion in the NNI budget supplements published by NSTC.
Therefore, we reviewed data on the individual projects included in the
agencies' total EHS research funding in 2010. We provided the agencies
with a spreadsheet template to use in listing their projects, which we
based on the spreadsheet OMB had used in its data calls to collect
project-level data from agencies on their 2006 and 2009 EHS research
projects. Consistent with the approach used in OMB's data calls, we did
not ask the agencies to use a standardized definition of what constitutes a
"project;" instead, we deferred to each agency to identify their projects
and consulted with agency officials as needed.

For some projects at NIH, NIOSH, and NIST, we grouped together some
of the agencies' data entries and counted them together for the purposes
of our analyses. Consequently, the numbers of projects we are reporting
may not match the number reported by these agencies elsewhere.
Specifically, for NIH, we grouped the 14 supplement grants together with
their main grants and counted each of those groups as a single project.
We also grouped together the individual subprojects of each of NIH's
multiproject "parent" grants reported as EHS and counted each of those
groups as a single project, for a total of seven groups of multiproject
grants. In addition, we grouped together two grants made by different NIH
institutes for the same project. For NIOSH, we identified two project
entries with identical project descriptions. After consulting with NIOSH
officials, we grouped those two entries together and counted them as a
single project. For NIST, the agency reported a separate entry for each of
its Project Tracking Numbers and explained that a single technical project
is often associated with more than one Project Tracking Number. NIST
officials instructed us to group together the entries with identical titles and
count each of those groups as a single project.

To assess the reliability of the agencies' fiscal year 2010 data, we sent
each of the seven selected agencies a set of questions regarding the data
and the information systems used to produce and store them. We also
reviewed related supporting documentation, such as user manuals and
data dictionaries for information systems, and, for some projects, copies
of the associated interagency agreements. We examined the data for
obvious errors, compared each agency's total fiscal year 2010 PCA 7

funding reported to us to the totals reported in the NNI Supplement to the
President's 2012 Budget, and consulted with agency officials to
understand the reasons for any differences. We determined that the data
were sufficiently reliable for the purposes of our analyses. However, we
did not attempt to verify the funding amounts reported for each project—
for example, by reviewing documentation of agencies' actual
expenditures for each project.

We reviewed the qualitative project data, including abstracts and project
reports, to assess whether the agencies had appropriately categorized
the research projects they reported as EHS research—that is, whether
the projects met the definition of program component area (PCA) 7. As
described in this report, the definition of PCA 7—research primarily
directed at understanding the EHS impacts of nanotechnology
development and corresponding risk assessment, risk management, and
methods for risk mitigation—is the only written guidance available to
agencies on how to report their nanotechnology research in this program
component area. Determining whether research meets this definition is an
inherently subjective process. Therefore, rather than definitively conclude
that any projects did not meet the definition, we assigned projects a
designation of "not clearly PCA 7" if we were ultimately unable to identify
any EHS research, or it was not clear that EHS was the primary focus of
the research. For example, some projects studied various
nanotechnology-enabled drugs, but it was not clear to what extent the
research was directed at the safety of the drugs versus their efficacy.
Some other projects appeared to be closely related to program
component areas other than PCA 7, such as PCA 4 (instrumentation
research, metrology, and standards for nanotechnology) or PCA 8
(education and societal dimensions), and it was not clear that PCA 7 was
the most appropriate category. To minimize bias and to ensure the
consistency of our evaluation, two analysts independently analyzed the
project data and used their professional judgment to assess whether each
project met the PCA 7 definition. The two analysts discussed those
projects where they did not agree on whether the research was primarily
directed at EHS and reached agreement. For categorization of projects
that appeared questionable to us, we generally asked agencies to explain
why they had reported those projects as EHS research. We reviewed
agencies' responses and modified our determinations for projects as
appropriate given the additional support provided by the agencies. For
two projects funded by NIST and 18 by NSF, the agencies had reported
only a portion of the total project funding under PCA 7. In those instances,
we assessed whether at least some of the project was primarily directed
at EHS. However, we did not verify that the funding amounts reported as

EHS research were correct because that information is, according to
these agencies' officials, based on the discretion of the knowledgeable
agency program staff. We also reviewed our 2008 report in this area,
which found that 22 of 119 fiscal year 2006 projects reported as EHS
research were miscategorized, and recommended that the Director of the
Office of Science and Technology Policy (OSTP), in consultation with the
Directors of the National Nanotechnology Coordinating Office and OMB,
provide better guidance to agencies on how to report nanotechnology
EHS research.[2] We asked OMB and OSTP to identify any guidance
provided to agencies on how to report their EHS research and to describe
any steps taken since our prior report to improve such guidance.

To identify the nanomaterials that the seven selected NNI member
agencies focused on in their EHS research in fiscal year 2010, we
reviewed qualitative project data—such as abstracts and progress
reports—for those projects we determined were primarily directed at EHS.
We also consulted with agency officials. One analyst identified, and a
second analyst verified, the nanomaterials in the project documentation.
We attempted to identify the nanomaterials that the projects actually
studied in fiscal year 2010. However, some projects are multiyear efforts.
The level of detail in the documentation we reviewed for each project
varied, and the documentation did not always indicate which materials
were studied in which years. For the purposes of our analysis, we
assumed that all nanomaterials identified for a project were studied in
2010, unless the documentation specified otherwise. It is possible that
some additional nanomaterials were studied by these projects but not
referenced in the documents we reviewed. It is also possible that some of
the nanomaterials referenced in project documents were not actually
studied until later years of a project. However, we do not have any
reasons to suspect that such variances would be significant or would
substantially change our findings.

We assigned each particular nanomaterial identified to one of five
categories: carbon-based nanomaterials; metal-based nanomaterials;
semiconductor-based nanomaterials; organic nanomaterials; and other
nanomaterials. In addition, some projects were not targeted to particular
nanomaterials. An analyst with previous academic and professional

---

[2]GAO, *Nanotechnology: Better Guidance Is Needed to Ensure Accurate Reporting of
Federal Research Focused on Environmental, Health, and Safety Risks,* GAO-08-402
(Washington, D.C.: Mar. 31, 2008).

experience in nanotechnology helped identify the categories we used and verified that the materials were placed in the appropriate categories for each project. After reviewing the fiscal year 2010 EHS research by the selected agencies, we selected the five categories of nanomaterials we used, based on the primary composition of the materials, to encompass the range of materials identified in the projects. To inform our selection of the categories we used for grouping nanomaterials, we reviewed literature related to nanotechnology EHS issues from federal agencies (NIH and EPA), nonprofit organizations (the Woodrow Wilson International Center for Scholars' Project on Emerging Nanotechnologies and the World Technology Evaluation Center, Inc.), and a market research firm that tracks nanotechnology (Lux Research). We found that various classes or categories were used to present information on nanomaterials. For materials identified as composites, we assigned them to categories based on the primary nanomaterials from which they were formed—for example, we assigned carbon nanotube nanocomposites to the carbon-based nanomaterials category.

To determine the extent to which the NNI member agencies collaborate with stakeholders on nanotechnology EHS research and related strategies, we (1) discussed with agency officials how their agencies collaborate on nanotechnology EHS research and NNI's role in facilitating that collaboration, and (2) we obtained documentation on these collaborative efforts. We conducted a review of formal collaborative efforts (i.e., those that are documented in written agreements) that focused on nanotechnology EHS research initiated from February 2008 to October 2011 while the 2008 NNI *Strategy for Nanotechnology-Related EHS Research* was in effect.[3] We chose this time frame because this strategy—the first NNI strategy to focus on EHS research—established collaboration as necessary for its implementation efforts. The strategy was in effect for a number of years, which allowed us to get a more complete picture of the number of collaborative projects completed during its implementation. We only included collaborative efforts that were formalized by written agreements because those provided a clearer description of the agencies' contributions of resources. We obtained an initial list of such efforts from published NNI sources as well as an OMB data call used to compile EHS-related nanotechnology projects for the

---

[3]NSTC, Committee on Technology, Subcommittee on Nanoscale Science, Engineering, and Technology, *National Nanotechnology Initiative Strategy for Nanotechnology-Related Environmental, Health, and Safety Research* (February 2008).

2011 NNI EHS research strategy. We sent our initial list to the selected
NNI agencies for their review and revision, and analyzed supporting
documents provided by the agencies. We provide relevant information
from the supporting documents for each of the projects within our scope
in appendix II.

We administered a web-based questionnaire to a nonprobability sample
of nonfederal stakeholders,[4] including those affiliated with academia,
companies, nongovernmental organizations (NGO), and state and local
governments, to obtain their views on collaboration with the NNI member
agencies on EHS research and the EHS research strategies. Because
the population of nonfederal stakeholders is diverse, wide-ranging, and
difficult to reliably define, we chose to use a nonprobability sample of
individuals who (1) had expertise in the field of nanotechnology, and (2)
had interacted with the NNI in the past few years, or who were
representatives of organizations and companies suggested to us through
our scoping interviews. We identified relevant potential respondents from
several sources, including the list of participants in NNI workshops that
were used to solicit input for the 2011 NNI EHS research strategy[5] and
participants in a review of the NNI by the President's Council of Advisors
on Science and Technology. We also included members of
nanotechnology-related trade associations and other organizations
through interviews with nanotechnology experts using an iterative
process, often referred to as "snowball sampling." We excluded federal
employees and contractors, individuals or organizations that are not
located in the United States, and those for whom we could not obtain
contact information. In addition, individuals who were not considered to
have expertise in the nanotechnology field based on their job titles were

---

[4]A nonprobability sample is a sample in which some items in the population have no
chance, or an unknown chance, of being selected. Results from nonprobability samples
cannot be used to make inferences about a population. However, information from the
sample provided illustrative examples.

[5]In 2009 and 2010, the NNI held four public workshops: (1) Human and Environmental
Exposure Assessment (February 2009); (2) Nanomaterials and the Environment &
Instrumentation, Metrology, and Analytical Methods (October 2009); (3) Nanomaterials
and Human Health & Instrumentation, Metrology, and Analytical Methods (November
2009); and (4) Risk Management Methods & Ethical, Legal, and Societal Implications of
Nanotechnology (March 2010).

excluded.[6] These sources provided us with a list of 228 stakeholders chosen as potential respondents. Of these 228 individuals, five subsequently were excluded from our sample because they had invalid e-mail addresses, were duplicates, or were not available for an extended period, giving us a final sample of 223 potential respondents. We categorized the population surveyed by affiliation type, usually based on information provided on each entity's website, as follows: (1) "academic," which includes universities or university-affiliated programs or centers; (2) "company," which includes companies, corporations, or other for-profit entities whose primary purpose is to sell products, services, or information; (3) "organization," which includes NGOs or other entities whose primary purpose is public interest or mutual benefit; and (4) "state and local government."

The questions included in the questionnaire were developed using issues raised in our scoping interviews and background literature search. Most of our questions provided respondents with answer options identified through our interviews and background literature searches. In certain cases, we also provided respondents with an "other" category which allowed them to write in additional responses we did not identify. However, because few respondents chose this option, we presented only some of these responses as illustrative examples. The questionnaire went through internal reviews, including independent GAO survey experts, as well as four pretests with two stakeholders who work for companies, one academic, and one stakeholder from an NGO. We chose the pretesters to reflect the diversity of affiliations of the nanotechnology experts included among our respondents. We revised and clarified the questions and introductory material to the questionnaire based on comments obtained in the internal reviews and from the pretesters.

We administered the questionnaire through a secure server. When we completed the final survey questions and format, we sent an e-mail announcement of the questionnaire to the nonfederal stakeholders in our sample on September 23, 2011. Stakeholders were notified that the questionnaire was available online and were given unique passwords and user names on October 11, 2011. We sent follow-up e-mail messages on

---

[6]For example, administrative assistants or staff managers were excluded because they are unlikely to have subject matter expertise. In another case, an individual in a consultancy that does not focus on nanotechnology was excluded, based on advice from the head of a nanotechnology trade association.

October 18, 2011, October 25, 2011, and November 2, 2011 to those who
had not yet responded. Then we contacted all nonrespondents by
telephone, starting on October 21, 2011. The questionnaire was available
online until December 6, 2011.

Because this was not a sample survey, it has no sampling errors.
However, the practical difficulties of conducting any survey may introduce
errors, commonly referred to as nonsampling errors. For example,
difficulties in interpreting a particular question, sources of information
available to respondents, or entering data into a database or analyzing
them can introduce unwanted variability into the survey results. We took
steps in developing the questionnaire, collecting the data, and analyzing
them to minimize such nonsampling errors. For example, we worked with
social science survey specialists to design the questionnaire. We
pretested the draft questionnaire with four nonfederal stakeholders to
ensure that the questions were relevant, clearly stated, and easy to
understand. When we analyzed the data, an independent analyst
checked all computer programs. Since this was a web-based
questionnaire, respondents entered their answers directly into the
electronic questionnaire, eliminating the need to key data into a database,
minimizing error.

Of the 223 potential respondents, 138 completed the questionnaire during
our time frame, for an overall response rate of about 62 percent.[7] The
numbers responding by affiliation type as categorized by us were
(1) "academic"—42 out of 62; (2) "company"—65 out of 114;
(3) "organization"—24 out of 39; and (4) "state and local government"—
7 out of 8.

To determine the extent the NNI strategy documents address desirable
characteristics of national strategies, we compared three key NNI
strategic documents[8] against criteria for desirable characteristics of

---

[7]Not all individuals responded to every question.

[8]These three strategic documents are: (1) NSTC, Committee on Technology,
Subcommittee on Nanoscale Science, Engineering, and Technology, *National
Nanotechnology Initiative Strategic Plan* (February 2011); (2) NSTC, Committee on
Technology, Subcommittee on Nanoscale Science, Engineering, and Technology,
National Nanotechnology Initiative Environmental, Health, and Safety Research Strategy
(October 2011); and (3) NSTC, Committee on Technology, Subcommittee on Nanoscale
Science, Engineering, and Technology, *The National Nanotechnology Initiative
Supplement to the President's 2012 Budget* (February 2011).

national strategies. Prior GAO reports have identified six desirable characteristics for national strategy documents that would help shape the policies, programs, priorities, resource allocations, and standards that would enable federal agencies and other stakeholders to implement the strategies and achieve the identified results.[9] National strategies that address these characteristics offer policymakers and implementing agencies a management tool that can help ensure accountability and more effective results.[10] Table 5 provides the desirable characteristics and the elements we looked for.

Table 5: Elements of Desirable Characteristics of National Strategies

Desirable characteristic	Brief description	Elements of characteristic
Purpose, scope, and methodology	Addresses why the strategy was produced, the scope of its coverage, and the process by which it was developed.	• Statement of broad or narrow purpose, as appropriate. • How it compares and contrasts with other national strategies. • What major functions, mission areas, or activities it covers. • Principles or theories that guided its development. • Impetus for strategy (e.g., statutory requirement or event). • Process to produce strategy (e.g., interagency task force; state, local, or private input). • Definition of key terms.
Problem definition and risk assessment	Addresses the particular national problems and threats the strategy is directed toward.	• Discussion or definition of problems, their causes, and operating environment. • Risk assessment, including an analysis of threats and vulnerabilities. • Quality of data available (e.g., constraints, deficiencies, and "unknowns").

[9]GAO, *National Capital Region: 2010 Strategic Plan is Generally Consistent with Characteristics of Effective Strategies,* GAO-12-276T (Washington, D.C.: Dec. 7, 2011); *Influenza Pandemic: Further Efforts Are Needed to Ensure Clearer Federal Leadership Roles and an Effective National Strategy,* GAO-07-781 (Washington, D.C.: Aug. 14, 2007); *Financial Literacy and Education Commission: Further Progress Needed to Ensure an Effective National Strategy,* GAO-07-100 (Washington, D.C.: Dec. 4, 2006); *Combating Terrorism: Evaluation of Selected Characteristics in National Strategies Related to Terrorism,* GAO-04-408T (Washington, D.C.: Feb. 3, 2004).

[10]GAO-07-781.

Desirable characteristic	Brief description	Elements of characteristic
Goals, subordinate objectives, activities, and performance measures	Addresses what the strategy is trying to achieve, steps to achieve those results, as well as the priorities, milestones, and performance measures to gauge results.	• Overall results desired (i.e., "end-state"). • Hierarchy of strategic goals and subordinate objectives. • Specific activities to achieve results. • Priorities, milestones, and outcome-related performance measures. • Specific performance measures. • Process for monitoring and reporting on progress. • Limitations on progress indicators.
Resources, investments, and risk management	Addresses what the strategy will cost, the sources and types of resources and investments needed, and where resources and investments should be targeted by balancing risk reductions and costs.	• Resources and investments associated with the strategy. • Types of resources required, such as budgetary, human capital, information technology, research and development, contracts. • Sources of resources (e.g., federal, state, local, and private). • Economic principles, such as balancing benefits and costs. • Resource allocation mechanisms, such as grants, in-kind services, loans, or user fees. • "Tools of government" (e.g., mandates or incentives to spur action). • Importance of fiscal discipline. • Linkage to other resource documents (e.g., federal budget). • Risk management principles.
Organizational roles, responsibilities, and coordination	Addresses who will be implementing the strategy, what their roles will be compared to others, and mechanisms for them to coordinate their efforts.	• Roles and responsibilities of specific federal agencies, departments, or offices. • Roles and responsibilities of state, local, private, and international sectors. • Lead, support, and partner roles and responsibilities. • Accountability and oversight framework. • Potential changes to current organizational structure. • Specific processes for coordination and collaboration. • How conflicts will be resolved.
Integration and implementation	Addresses how a national strategy relates to other strategies' goals, objectives, and activities—and to subordinate levels of government and their plans to implement the strategy.	• Integration with other national strategies (horizontal). • Integration with relevant documents from implementing organizations (vertical). • Details on specific federal, state, local, or private strategies and plans. • Implementation guidance. • Details on subordinate strategies and plans for implementation (e.g., human capital, and enterprise architecture).

Source: GAO-04-408T.

To assess whether the documents addressed these desirable
characteristics, two analysts independently reviewed the three NNI
strategy documents for the elements of each characteristic. We examined
the documents for inclusion of each element and for sufficient specificity
and detail. The two analysts compared their assessments, discussed any
discrepancies, and agreed upon a determination for each element. Each
characteristic was then given a rating of "address," "partially address," or
"does not address." According to our methodology, the strategy
documents "address" a characteristic when it explicitly includes all, or
nearly all, elements of the characteristic and has sufficient specificity and
detail. The documents "partially address" a characteristic when it includes
some or most of the elements with sufficient specificity and detail. The
"partially address" category includes a range that varies from explicitly
including most of the elements to including as few as one of the elements
of a characteristic. A strategy "does not address" a characteristic when it
does not include any elements of a characteristic or references are too
vague or general to be useful.

Because the three NNI documents have different purposes and
audiences, an individual document need not address each element of
each desirable national strategy characteristic. Therefore, we reviewed
the three NNI strategy documents as a whole. In the case of the 2011
NNI strategic plan and Supplement to the President's 2012 Budget, we
focused on sections related to EHS research, where applicable.

We identified agency documents related to nanotechnology EHS
research strategies and performance. Documents were identified based
on interviews with agency officials, review of agency websites, and a data
collection instrument sent to agencies requesting any documentation
related to the planning or performance of their nanotechnology EHS
research programs. In order to assess the extent of linkage between
agencies' nanotechnology EHS research and national priorities of the
NNI, we evaluated agency management documents for explicit referrals
to NNI activities or guidance, including strategic documents such as the
NNI strategic plan or EHS research strategy, or activities of the NNI's
planning and coordinating bodies. We also reviewed agencies'
Performance Reports in order to assess agencies' use of performance
information related to nanotechnology EHS research efforts.

We conducted this performance audit from February 2011 through May 2012 in accordance with generally accepted government auditing standards. Those standards require that we plan and perform the audit to obtain sufficient, appropriate evidence to provide a reasonable basis for our findings and conclusions based on our audit objectives. We believe that the evidence obtained provides a reasonable basis for our findings and conclusions based on our audit objectives.

Appendix II: Collaborative Nanotechnology Environmental, Health, and Safety Research Agreements

Table 6 identifies collaborative nanotechnology EHS research project agreements initiated by the NNI member agencies included in our review from February 2008 to October 2011. The table identifies only those projects specifically focused on nanotechnology. All the information in the table is based on the written agreements, which were provided by NNI member agencies' officials.

Table 6: Collaborative Nanotechnology Environmental, Health, and Safety Research Agreements

Participating federal agencies	Nonfederal partners	Date initiated	Type of collaboration	Title	Purpose
Consumer Product Safety Commission (CPSC), National Institute of Standards and Technology (NIST)	NA	April 2008	Memorandum of understanding	Memorandum of understanding between CPSC and NIST's Building and Fire Research Laboratory	Develop best practices to identify the presence of and characterize and quantify potential release of nanomaterials from flame retardant consumer products, to help assess the safety of these products. The cooperating agencies plan to work closely to expedite the development and availability of methods for the characterization and quantification of the release of nanomaterials from a variety of products (e.g., textiles, plastics) which are assessed using any of a number of standard fire tests.
CPSC, National Institute for Occupational Safety and Health (NIOSH)	NA	July 2008	Interagency agreement	Nano product exposure assessment	Conduct laboratory investigations on the emissions of nanoscale titanium dioxide from selected consumer products. The project also is to involve developing the appropriate sampling and analysis methods for nanomaterials incorporated into selected consumer products. The first products to be investigated are bathroom cleaners that utilize titanium dioxide as a catalyst to destroy bacteria or to break down organic material. These products are believed to be used by consumers and in occupational settings (e.g., hotel cleaning). CPSC staff interests involve emissions during consumer use while NIOSH is concerned with emissions that occur in occupational use scenarios.
CPSC, NIST	NA	Initiated July 2009, extended September 2009	Interagency agreement	Exposure and fire hazard assessment of nanoparticles in fire safe consumer products	Determine if nanoparticles that are traditionally used to improve the fire performance of foam and barrier fabrics are released during simulated normal conditions (i.e., mechanical compression and saliva). In addition, determine if the release of nanoparticles also creates a decrease in the fire performance of these components.
CPSC, NIOSH	NA	March 2010	Interagency agreement	Pulmonary effects of titanium dioxide nanoparticles released from aerosol spray products	Evaluate the acute bioactivity of a particulate aerosol from a bathroom cleaner or sanitizer containing titanium dioxide nanoparticles by exposing rats by inhalation to the aerosolized spray particles (low and high dose) for 2 hours and monitor pulmonary responses at 1, 7, and 28 days post-exposure.

Participating federal agencies	Nonfederal partners	Date initiated	Type of collaboration	Title	Purpose
CPSC, NIOSH	NA	April 2010	Interagency agreement	Exposure assessment of silver nanoparticles in treated products	Use scientifically credible protocols to evaluate exposure potential to nanosilver from related products. NIOSH plans to test the broad hypothesis that nanosilver is released from products under conditions of intended use.
CPSC, EPA	NA	May 2010	Interagency agreement	Exposure assessment of silver nanoparticles in select children's consumer products	Evaluate children's potential exposures to nanosilver from consumer products marketed exclusively for children (e.g., pacifiers, plush toys) using scientifically credible protocols. EPA plans to test the broad hypothesis that nanosilver is released from children's consumer products under conditions of intended use.
CPSC, NIOSH	NA	March 2011	Interagency agreement	Simulated worker exposure to silver from use of nano-enabled consumer products	Use scientifically defensible methods to evaluate the release of silver from consumer products and model exposure potential. NIOSH plans to test the broad hypothesis that silver is released from nanosilver enabled products under conditions of intended use in the work environment.
CPSC, NIOSH	NA	May 2011	Interagency agreement	Exposure assessment and pulmonary effects of silver nanoparticles released from aerosol spray products	Characterize the generated particulate from a product containing silver nanoparticles aerosol including the mass, particle size distribution, and chemical composition of the particles. NIOSH staff plan to expose rats by inhalation to the aerosolized spray particles (low and high dose) for 2 hours and monitor pulmonary responses at 1, 7, and 28 days post-exposure.
CPSC, NIST	NA	May 2011	Interagency agreement	Exposure assessment of carbon nanotubes in sports equipment	Determine if nanomaterials that are reported to be used in selected sports equipment are present in these products and are released during simulated use conditions. In addition, this project seeks to determine if the release of nanoparticles improve the performance of these products.
CPSC, NIST	NA	July 2011	Interagency agreement	Quantify and characterize released nanoparticles from thermoset and thermoplastic samples	Measure the impact of nanotechnologies on the flammability of the thermoplastic- and thermoset-based consumer products and develop risk models for human exposure to materials released from these consumer products.

Participating federal agencies	Nonfederal partners	Date initiated	Type of collaboration	Title	Purpose
CPSC, NIST	NA	August 2011	Interagency agreement	Characterization of airborne nanoparticles released from consumer products	Develop testing and measurement protocols for determining the quantities and properties of nanoparticles released from flooring finishes and interior paints, and their subsequent airborne concentrations. Develop essential methods and data to assess the release as well as accumulation of nanoparticles at the surfaces of these products that seek to assist in estimating occupant exposure and developing strategies to manage and mitigate these exposures.
CPSC, Environmental Protection Agency (EPA)[a]	NA	September 2011	Interagency agreement	Risk assessment for manufactured nanoparticles used in consumer products	Determine whether the nanoparticles exposures to humans and organisms resulting from consumer product and environmental exposures will substantially differ in both physicochemical and toxicological properties from nanoparticles that are newly synthesized in the laboratory or incorporated directly into a product. These differences may have significant consequences with respect to nanoparticles bioavailability and other toxicity measures. This study plans to test these measures in *in vitro* and *in vivo* experiments, and produce mechanism-based results relating toxic effects to nanoparticles physicochemical properties.
CPSC, National Science Foundation (NSF)	NA	September 2011	Interagency agreement	Interagency agreement between the CPSC and NSF	Develop innovative tools for measuring the potential health impact of nanotechnologies used in the flammability treatment of thermoplastic- and thermoset-based consumer products and develop risk models for human exposure to materials released from these consumer products.
EPA	United Kingdom (UK) Natural Environment Research Council, UK Engineering and Physical Sciences Research Council, UK Department of Environment, Food, and Rural Affairs, UK Environment Agency	March 2009	Memorandum of understanding	Memorandum of understanding for a joint request for applications for nanotechnology research proposals	Establish a cooperative framework for the EPA, the UK Natural Environment Research Council, the UK Engineering and Physical Sciences Research Council, the UK Department of Environment, Food, and Rural Affairs, and the UK Environment Agency to organize, plan, and execute a joint Request for Applications for research related to the environmental and health implications of released nanomaterials.

Participating federal agencies	Nonfederal partners	Date initiated	Type of collaboration	Title	Purpose
EPA, NSF, U.S. Department of Agriculture's National Institute of Food and Agriculture	NA	Solicitation issued November 2009	Grant or cooperative agreement	Increasing scientific data on the fate, transport, and movement and behavior of engineered nanomaterials in selected environmental and biological matrices	Increase scientific knowledge on the partitioning of nanomaterials in various media and increase data on movement and transformation capacities. Address the urgent needs to scientifically understand the fate and properties of nanoscale materials and additives that may be used or introduced into foods. Assess the adequacy of the existing characterization methods to study the critical questions, and establish the baseline for the needs of new characterization methodology. Provide guidance on the extent of future investigation needs on the nanoscale food materials and additives.
EPA, NIST	NA	July 2011	Interagency agreement	Measurements to support research in nanoparticles	Investigate aggregation of engineered nanoparticles in an artifact of air sampling on filter media and investigate how engineered nanoparticles change microscopically with age in the atmosphere.
EPA	International Life Sciences Institute Research Foundation[b]	September 2011	Contract	Support for better understanding of nanomaterials to the environment workshop (Phase 2)	Evaluate the "state of the science" for release measurement for multi-walled carbon nanotubes in polymer. The project seeks to use information (methods, studies, guidance, etc.) collected in response to a data call and literature search and provide that information to groups of subject matter experts selected by the steering committee. This information is to also be made available in the project's information catalogue to the extent it does not reveal confidential business information.
Food and Drug Administration (FDA), National Institutes of Health's (NIH) National Cancer Institute (NCI), and NIST	NA	July 2009	Memorandum of understanding	Memorandum of understanding between the FDA, NCI, and NIST for the Nanotechnology Characterization Laboratory (NCL) and related nanotechnology activities	Facilitate the development of nanotechnologies that constitute novel research tools and safer, more effective cancer therapies by establishing a framework for effective risk identification, assessment, and evaluation of emerging products based on nanotechnology. This collaboration is to be focused primarily on the NCL and NIST directly related activities.

Participating federal agencies	Nonfederal partners	Date initiated	Type of collaboration	Title	Purpose
FDA's Center for Biologics Evaluation and Research, NCL, NIST	Science Applications International Corporation (SAIC)-Fredrick[c]	February 2011	Grant mechanism or other opportunities	*In vitro* evaluation of effects of nanomaterials on blood platelets	Develop a panel of *in vitro* assays for evaluation of effects of nanomaterials on blood platelets. Research validation of the assays for different representative types of nanomaterials, soluble nanoparticles, and nanoobjects in suspension—representative metal, carbon, polymer, dendrimer, liposome nanoparticles, nanoparticles containing composites, or nanostructured macrosurfaces. Investigate a structure-function relationship of a nanoparticle-platelet interaction in a model of different size-charge modified poly(amidoamine) dendrimers.
FDA's National Center for Toxicological Research, Center for Food Safety and Applied Nutrition, Center for Veterinary Medicine; National Toxicology Program; Pacific Northwest National Laboratories	NA	February 2011	Grant mechanism or other opportunities	Development and evaluation of exposure dosimetry methods to optimize the standard *in vitro* mammalian genotoxicity assays for assessing engineered nanomaterials	Evaluate whether the existing *in vitro* mammalian genotoxicity assay is suitable for assessing the genotoxicity of nanomaterials. Explore the possible mechanisms underlying genotoxicity of engineered nanomaterials by conducting genomic analysis. Identify potential improvements to the assay and general strategies for evaluating nanomaterials. Examine whether the suitable methods and other experiences learned from the micronucleus assay are applicable to other genotoxicity tests, such as mouse lymphoma assay and *in vivo* micronucleus assay.
FDA, NCI, NCL	SAIC-Fredrick[c]	July, 2011	Interagency agreement	*In vitro* dermal penetration and *in vivo* distribution of nanoparticles used in food and cosmetics	Utilize free and surface model nanoparticles (such as dendrimers, nano-silver or nano-gold) with the size range typically found in food and cosmetics to assess their characteristics and correlate the effect of surface modification on the ability of nanomaterials to penetrate various tissue barriers in different *in vitro* and *in vivo* systems. Supply information related to how nanomaterial physiochemical properties affect the ability of these nanoscale materials to interact with various biological systems and, as a result, to evaluate the benefits versus the risks when making regulatory approval decisions in order to facilitate the development and commercialization of safe nano-based cosmetics, foods, and other FDA-regulated products.
FDA's Center for Food Safety and Applied Nutrition and National Center for Toxicological Research	Key Laboratory for Biomedical Effects of Nanomaterials and Nanosafety, Chinese Academy of Sciences	Fiscal year 2011	Grant mechanism or other opportunities	Use of electron spin resonance spectroscopy and biomarkers of oxidative damage to assess the safety of nanoscale materials used in cosmetics	Develop rapid and predictive tests to assess toxicity elicited by nanosized materials. The tests may be used by the agency and the regulated industry to identify potentially hazardous nanosized materials in cosmetic products.

Participating federal agencies	Nonfederal partners	Date initiated	Type of collaboration	Title	Purpose
NIH's National Institute of Environmental Health Sciences (NIEHS), NIOSH	NA	Initiated in June 2008 and extended in June 2009 and March 2010	Interagency agreement	Comprehensive toxicological assessment of occupationally-relevant exposures	Characterize workplace exposure to selected engineered nanoparticles. This proposal is to focus on addressing the lack of exposure data. Twelve field investigations at sites where engineered nanoparticles are handled and manufactured are to be conducted. The focus of these investigations is to be nanoparticles of interest to NIEHS and NIOSH.
NIEHS, NIOSH	NA	Initiated in February 2009 and extended in December 2009	Interagency agreement	Cardiovascular toxicity assessment of subchronic inhalation exposure to fullerene C60	Evaluate potential cardiovascular toxicity of subchronic inhalation exposure to fullerene C60 in animal models using molecular and biochemical analysis of cardiovascular tissue and blood samples. The findings are to be correlated with the histopathological, particle distribution, and blood chemistry results being obtained in other National Toxicology Program studies of this proposal.
NCI, NIH's National Institute of Biomedical Imaging and Bioengineering, NIEHS	NA	Initiated in April 2010 with NIEHS; NCI joined in September 2010	Direct citation agreement	Nano registry	Establish an authoritative nanomaterial registry, whose primary function is to provide consistent and authoritative information on the biological and environmental interactions of well-characterized nanomaterials, as well as links to associated publications, modeling tools, computational results and manufacturing guidance. The registry seeks to provide a curated data source and assessment information on the health and environmental effects of well-characterized nanomaterials. A set of minimal information about nanomaterials, ontology, and standards are to be developed through a community effort with broad representation to establish the authoritativeness of the registry.
NCI, NCL, NIEHS	SAIC-Fredrick[c]	April 2010	Direct citation agreement	NIEHS nanomaterial characterization and informatics	NIEHS requests NCI and SAIC-Fredrick support for physiochemical characterization of engineered nanomaterials. NCL seeks to provide data to support and compare in-house characterization from program grantees. Facilitate the dissemination of findings and provide maximum impact to the field by providing direct access to a database for inclusion of nanomaterials physiochemical characterization and biological response data.

Participating federal agencies	Nonfederal partners	Date initiated	Type of collaboration	Title	Purpose
NIEHS, NIOSH	NA	November 2010	Interagency agreement	Comprehensive toxicological assessment of occupationally-relevant exposures	Identify and assess workplaces where select engineered nanomaterials are manufactured or used and characterize potential worker exposures to those nanomaterials. Address the lack of information available on the identity and location of manufacture or use of engineered nanomaterials with high commercial potential and the lack of worker exposure data. Conduct 12 field investigations at sites where engineered nanomaterials are handled and manufactured. The focus of these investigations is to be on nanomaterials of interest to NIEHS and NIOSH. Give priority to materials on the Organisation for Economic Co-operation and Development's list of representative manufactured nanomaterials as well as those that are receiving increased attention due to requests to manufacture, use, or distribute in the United States.
NIEHS, NIOSH	NA	March 2011	Interagency agreement	Durability of nanoscale cellulose fibers in artificial human lung fluids	Characterize the physicochemical properties and dissolution behaviors of nanoscale cellulose fiber materials in artificial lung fluids to determine their biodurability.
NIOSH	Institute of Occupational Medicine[d]	March 2009	Collaborative research	Risk assessment of engineered nanoparticles	Conduct short-term inhalation experiments in mice to investigate short-term kinetics and health effects (e.g., pulmonary, cardiovascular, and central nervous system responses) of one or two selected engineered nanoparticles. Develop risk assessment methods, including biomathematical models of exposure-dose response, sensitivity and uncertainty analyses, and extrapolation from in vivo studies in rodents to predict internal dose and risk in humans. Provide exposure data in workplaces where nanoparticles are produced or used and develop risk management strategies.
NIOSH	University of Massachusetts, Lowell	June 2009	Contract	Professional services for developing a guidance document: Safe Practices for Working with Engineered Nanomaterials in Research Laboratories	Review the available scientific literature, conduct empirical measurements in laboratories, and apply best professional judgment to the development of a draft guidance document on safe practices for engineered nanomaterials in research laboratories.

Participating federal agencies	Nonfederal partners	Date initiated	Type of collaboration	Title	Purpose
NIOSH	Rice University's International Council on Nanotechnology (ICON)	August 2009	Contract	Development and deployment of a wiki to share good risk management practices for engineered nanomaterials	Develop a workspace for good risk management practice information, housed on the ICON GoodNanoGuide wiki. Using wiki technology for sharing risk management information, successful practices, and case studies, seek to provide NIOSH stakeholders and the public with an easy to use, familiar, flexible, and comprehensive platform to obtain critical occupational safety and health information. Discussion forums can be established for users to collaborate on a wide range of topics, including novel facility design solutions, improved control technologies, or creative risk characterization techniques. New users can engage the experienced user community for nanomaterial management advice. Developing a wiki-based information sharing platform through ICON has the potential for reducing barriers to providing information to a government site.
NIOSH, Department of Defense (DOD)	NA	October 2009	Memorandum	Examination of potential occupational exposures of engineered nanomaterials at the DOD: a preliminary framework for DOD and NIOSH	Evaluate environment, safety, and occupational health issues related to the creation, handling, and use of engineered nanomaterials at facilities across all DOD agencies. This partnership can be effected across the DOD or on an individual basis (Department of Army, Department of Navy, etc.), dependent upon the degree of interest and need for such environmental, safety, and occupational health evaluations.
NIOSH	Battelle Memorial Institute Pacific Northwest Division[e]	March 2010	Memorandum of understanding	Memorandum of understanding between Battelle Memorial Institute Pacific Northwest Division and NIOSH	Combine the capabilities of NIOSH and Battelle to advance the fundamental science supporting the assessment of nanotechnologies. Specifically, NIOSH and Battelle seek to engage in a joint research program studying the comparative toxicity of commercially important nanomaterials (e.g., single-walled carbon nanotubes and silica).
NIOSH	Center for High-Rate Nanomanufacturing at the University of Massachusetts, Lowell	August 2010	Contract	Engineering controls for nanomaterial handling	Use data from field research to evaluate the performance of available engineering controls and promote their use for nanomaterial manufacturing and handling. Develop guidance for safely handling nanomaterials.

Participating federal agencies	Nonfederal partners	Date initiated	Type of collaboration	Title	Purpose
NIOSH	ICON	August 2010	Contract	Develop and deploy a wiki to share good risk management practices for engineered nanomaterials; follow-on deployment and evaluation	Perform follow-on deployment and evaluation tasks of work from previously completed contract. The objective of the platform, now known as the GoodNanoGuide, is to contribute to safety and health excellence by creating a generally accessible repository of critical safety and health research findings and best practice information to assist users in developing and administering effective safety and health programs to be used during the research, production, and use of engineered nanomaterials. GoodNanoGuide content is being contributed from a variety of sources, including private, government, and academic safety, health, and environmental practitioners and from the various national and international consortia that have been created to foster good practices. The long-range intent is to have a virtual warehouse (or library) where health, safety, and environmental professionals and the public can go to share information and collaborate on particular topics.
NIOSH, Occupational Safety and Health Administration	NA	August 2010	Interagency agreement	Guidance for nanomaterials workers	Investigate and characterize potential hazards associated with manufactured nanomaterials, with the goals of developing and recommending occupational safety and health measures to protect workers who manufacture and handle these materials. Build upon ongoing collaboration to investigate a control-focused guidance strategy for assisting small business employers in identifying and controlling workplace chemical hazards.
NIOSH	Lovelace Respiratory Research Institute	September 2010	Contract	Deposition of carbon nanotubes in a human respiratory tract replica	Obtain measurement data on the deposition efficiency and size distribution of multi-walled carbon nanotubes and single-walled carbon nanotubes in a human respiratory tract replica. These data are to be used by NIOSH to calibrate and evaluate a preliminary biomathematical model to predict the deposition of airborne carbon nanotubes in the human respiratory tract. A key element of this research is to develop an approach to relate the carbon nanotubes particle morphology to their aerodynamic and diffusion diameters.

Participating federal agencies	Nonfederal partners	Date initiated	Type of collaboration	Title	Purpose
NIOSH	University of Massachusetts, Lowell, the Center for High-rate Nanomanufacturing	September 2010	Memorandum of understanding	NIOSH-Center for High-rate Nanomanufacturing partnership to advance research and guidance for occupational safety and health in nanotechnology	Advance workplace health and safety standards and practices and strengthen the resulting nanotechnology workforce. The Center for High-rate Nanomanufacturing, as a part of this partnership, seeks to collaborate with NIOSH to provide a state-of-the-art platform for the discovery and dissemination of fundamental knowledge in the emerging interdisciplinary fields of nanotechnology, including nanoscience, nanoengineering, and nanobiotechnology for the purpose of jointly developing effective occupational safety and health guidance.
NIOSH	College of Nanoscale Science and Engineering of the University at Albany—State University of New York	October 2010	Memorandum of understanding	NIOSH-College of Nanoscale Science and Engineering of the University at Albany partnership to advance research and guidance for occupational safety and health in nanotechnology	Establish a world-class intellectual and physical platform for occupational safety and health research, development, educational, and business initiatives leading to the discovery, demonstration, and deployment of new guidelines, recommendations, and findings relating to the potential human health impacts of exposure to engineered nanomaterials.
NIOSH	Applied Research Associates, Inc.[9]	March 2011	Contract	Development of a preliminary biomathematical model to predict deposition of carbon nanotubes and nanofibers in the respiratory tract of workers	Extend the carbon nanotubes deposition model to accommodate additional input data on both aerodynamic and diffusion diameters, and on particle number concentration as well as mass concentration. An acceptable extended model is to provide predictions of the deposition of carbon nanotubes of different sizes and shapes (i.e., single-walled carbon nanotubes and multi-walled carbon nanotubes) in the human respiratory tract. The model is to be calibrated and evaluated—evaluation includes determining the particle properties and model parameters that provide the best predictions of multi-walled carbon nanotubes and single-walled carbon nanotubes deposition in the human respiratory tract replica. The carbon nanotubes deposition model is to be linked with an initial model to predict carbon nanotubes clearance and retention in the respiratory tract.
NIOSH	ICON	August 2011	Contract	Develop content and evaluate dissemination through the GoodNanoGuide wiki	Develop focused content that includes new work practices and case studies, moderating the wiki, providing user support, and conducting an evaluation of the effectiveness of the GoodNanoGuide.

Participating federal agencies	Nonfederal partners	Date initiated	Type of collaboration	Title	Purpose
NIOSH, NIST, DOD, Department of Energy	Over 20 private sector entities	Program first initiated in July 2005; multiple site visits occurred between February 2008 and October 2011	Letter of introduction	Site visits by NIOSH for the purpose of evaluating engineered nanomaterials in the workplace	Enter nanomaterial manufacturing facilities to characterize process and evaluate the potential for worker exposure to nanoparticles. NIOSH may publish information from the overall study in the primary literature, and may supplement its *Nanotechnology* document on its website to incorporate, in a de-identified manner, observations or findings specific to work practices, exposure controls, or risk management practices that come from the visit that may be of benefit to the rest of the nanomaterial manufacturing community.
NIST, NIOSH, NIH	Agreement with International Alliance for NanoEHS Harmonization;[h] other partners for the alliance also include over 10 private entities and international organizations; Collaborators include the European Commission's Institute for Reference Materials and Measurements	October 2008	Participation agreement	International Alliance for NanoEHS Harmonization	Create protocols for a limited number of *in vitro* and *in vivo* toxicology tests on a small number of representative nanoparticles to enable a "round robin" study with the goal to achieve inter-laboratory reproducibility of results and verify and/or validate predictability of *in vitro* assays for *in vivo* responses. This is to involve at least the use of nanomaterials and biologicals from a single source, and a single set of protocols. The implementation of a round robin laboratory set of tests based on these protocols. Ongoing and further development of protocols that take into consideration study design and methodology, nanomaterial physicochemical properties, and quantitative structure function relationships at the nanobio interface for *in vitro* and *in vivo* protocol development. A next phase is to involve round robin studies of common nanomaterials and biological systems, where independent laboratories are to choose a different set of agreed approaches to interrogate the biological end-points.

Source: GAO analysis based on formal collaborative EHS research project agreements provided by NNI member agencies' officials.

[a]This agreement funded a grant selected through the cooperative framework signed between EPA and the UK.

[b]The International Life Sciences Institute Research Foundation is a nonprofit organization based in Washington, D.C. that focuses on advancing the methods and application of science in risk assessment, human nutrition, and the prevention of obesity.

[c]SAIC-Frederick is a wholly owned subsidiary of Science Applications International Corporation and operates exclusively under a single, long-term contract to the National Cancer Institute.

[d]The Institute of Occupational Medicine is an independent organization based in the UK that produces research, consulting, and other services related to occupational and environmental health, hygiene, and safety.

[e]Battelle Memorial Institute is an independent research and development organization based in Columbus, Ohio.

[f]The Lovelace Respiratory Research Institute is a private biomedical research organization dedicated to improving public health through research on the prevention, treatment, and cure of respiratory disease.

[g]Applied Research Associates, Inc. is a research and engineering company based in Albuquerque, New Mexico.

[h]International Alliance for NanoEHS Harmonization consists of internationally recognized experts drawn from academia and other scientific institutions from around the world. It is currently chaired by the Centre for BioNano Interactions at University College, Dublin.

Appendix III: GAO Contact and Staff Acknowledgments

GAO Contact	Frank Rusco, (202) 512-3841 or ruscof@gao.gov
Staff Acknowledgments	In addition to the contact named above, Dan Haas (Assistant Director), Krista Anderson, Nirmal Chaudhary, Elizabeth Curda, Lorraine Ettaro, Alison O'Neill, Tind Shepper Ryen, Jeanette Soares, Ruth Solomon, Hai Tran, and Jack Wang made key contributions to this report.

GAO's Mission	The Government Accountability Office, the audit, evaluation, and investigative arm of Congress, exists to support Congress in meeting its constitutional responsibilities and to help improve the performance and accountability of the federal government for the American people. GAO examines the use of public funds; evaluates federal programs and policies; and provides analyses, recommendations, and other assistance to help Congress make informed oversight, policy, and funding decisions. GAO's commitment to good government is reflected in its core values of accountability, integrity, and reliability.
Obtaining Copies of GAO Reports and Testimony	The fastest and easiest way to obtain copies of GAO documents at no cost is through GAO's website (www.gao.gov). Each weekday afternoon, GAO posts on its website newly released reports, testimony, and correspondence. To have GAO e-mail you a list of newly posted products, go to www.gao.gov and select "E-mail Updates."
Order by Phone	The price of each GAO publication reflects GAO's actual cost of production and distribution and depends on the number of pages in the publication and whether the publication is printed in color or black and white. Pricing and ordering information is posted on GAO's website, http://www.gao.gov/ordering.htm. Place orders by calling (202) 512-6000, toll free (866) 801-7077, or TDD (202) 512-2537. Orders may be paid for using American Express, Discover Card, MasterCard, Visa, check, or money order. Call for additional information.
Connect with GAO	Connect with GAO on Facebook, Flickr, Twitter, and YouTube. Subscribe to our RSS Feeds or E-mail Updates. Listen to our Podcasts. Visit GAO on the web at www.gao.gov.
To Report Fraud, Waste, and Abuse in Federal Programs	Contact: Website: www.gao.gov/fraudnet/fraudnet.htm E-mail: fraudnet@gao.gov Automated answering system: (800) 424-5454 or (202) 512-7470
Congressional Relations	Katherine Siggerud, Managing Director, siggerudk@gao.gov, (202) 512-4400, U.S. Government Accountability Office, 441 G Street NW, Room 7125, Washington, DC 20548
Public Affairs	Chuck Young, Managing Director, youngc1@gao.gov, (202) 512-4800 U.S. Government Accountability Office, 441 G Street NW, Room 7149 Washington, DC 20548

Please Print on Recycled Paper.

www.ingramcontent.com/pod-product-compliance
Lightning Source LLC
Chambersburg PA
CBHW081555170526

45166CB00009B/2705